Newton versus Relativity

Newton versus Relativity

Jean-Michel Rocard

Translated from a revised version of
Newton et la relativité
by Charles Billet and Jean-Michel Rocard

VANTAGE PRESS
New York

Copyright © 1992 by Jean-Michel Rocard

Published by Vantage Press, Inc.
516 West 34th Street, New York, New York 10001

Manufactured in the United States of America
ISBN: 0-533-09637-5

Library of Congress Catalog Card No.: 91-92014

0 9 8 7 6 5 4 3 2 1

Nature and nature's laws lay hid in night;
God said: "Let Newton be," and all was light!
—Alexander Pope

Contents

Introduction

Automobiles speed down our roads. Airplanes fly high over our heads. Satellites sweep across the sky while the stars rise and set with precise regularity. Who or what is responsible for these movements? Is there one single underlying cause? Indeed, does there have to be a cause?

In studying the movement of objects, the systematic description of observations made with respect to a given reference system is called *kinematics*. This is the description of different motions with no consideration as to their causes, meaning the forces acting upon these objects.

It is Newton's laws of motion as well as the understanding of the basic forces in nature that allow us to describe and predict motion—even the complex motions of flying, falling, vibrating, or exploding bodies. These laws of motion go further, for they enable us to discover properties that remain invariable and quantities that are conserved when all the rest undergoes change. In particular, starting with Newton's three laws of motion we may deduce the three laws of conservation governing an isolated system: (1) conservation of linear momentum, (2) conservation of angular momentum, and (3) conservation of energy.

These three laws of conservation can be applied to all areas of classical, relativistic, and quantum physics, to such an extent that when difficulties arise (for instance, when discrepancies turn up in balances applying to energy and linear and angular momentum) it becomes preferable to ''invent'' new particles such as photons or neutrinos rather than admit that the laws of conservation have been broken.

Many scientists, on the other hand, readily assert that Newton's third law, which states that action and reaction are always equal and opposite, may, indeed, be broken with no undue difficulty. The reason, they argue, is that since Newton's time science has discovered electromagnetism, special relativity, quantum mechanics, and other such niceties. And yet these assertions are often the source of confusion. The trouble becomes especially apparent in other topics, such as *inertial forces*—fictitious for some scientists and very real for others, even when the accelerated reference frame is devoid of all matter. There is also confusion in the classical problem of the rocket.

The purpose of this book is twofold: first, to demonstrate that the three laws of Newtonian motion are intimately connected with the three laws of conservation; second, to show that if the former cannot be broken, the latter cannot either.

In the first chapter we review the original statement of Newton's three laws of motion and then go on to show that, in light of the interactions between bodies through forces and couples, these statements need to be updated or generalized. Such generalized statements are in perfect keeping with the laws of conservation referred to above.

In the three chapters that follow we deal with three important topics: (1) variable mass systems, (2) inertial and Coriolis forces, and (3) Newton's third law (generalized). The study of these three subjects will lead us to conclude that there exists a certain analogy in the behavior of the three types of forces under consideration: gravitational attractive forces, inertial forces, and electromagnetic forces.

Chapter 5 is devoted to the relationship that may exist between Newton's laws and special relativity (and/or general relativity). Using a simple theory—the similarity between the gravitational and the electromagnetic interactions—and Newton's second law, we suggest a formula making it possible to calculate the advance of the perihelion of Mercury, Venus, the Earth, and Mars.

In chapter 6, we review a few applications of the topics set forth in the previous chapters.

These six chapters* are independent of each other and can thus be read separately in any order. However they may be read, I hope that the reader will have grasped what I set out to show: *that even if Einstein is right, Newton is not, for that reason, necessarily wrong.*

*Numbers in parentheses in the text refer to the references listed at the end of the book. Letters in bold characters, such as **v**, describe vector quantities.

Newton versus Relativity

Chapter 1
Newton's Laws of Motion

Classical mechanics deals with the motion of material bodies moving at velocities much lower than the speed of light (c). Quantum mechanics, on the other hand, is concerned with describing the behavior of atoms and subparticles, while relativistic mechanics includes particles moving at speeds close to that of light.

Classical mechanics—also called Newtonian mechanics—serves as a basis for many branches of physics, including astronomy, and is therefore of obvious importance. The predominance of Newton's three laws of motion is thus clearly understandable. The theory of classical mechanics dates back more than a thousand years, but the science as we know it today is the result of unending research and experiments conducted by curious men seeking to find explanations for the phenomena and the motions of objects they have observed—and still observe—going on around them or at great distances from them.

There is a tendency to limit Newton's laws of motion to classical physics. Yet we must point out that the classification of classical, quantum and relativistic physics is an arbitrary one. Nature and the universe do not make such distinctions when it comes to visible phenomena. What is more, as we shall see in chapter 5, these laws may be extended to the area of relativistic physics.

A Review of Newton's Laws of Motion

In this section we shall review the three laws of motion as they were formulated by Newton and as they have since been interpreted by the great majority of scientists.

Newton's First Law or the Law of Inertia

In its original form this law states that in the absence of any outside influence, a body at rest remains at rest, and a body in uniform motion remains in uniform motion (constant speed).

This statement has met with general acceptance and means that a free particle, i.e., one isolated in space, will remain at rest or move with constant speed, depending upon the initial conditions. And since states of rest or of uniform motion must be observed with respect to a reference system, the above statement of the law of inertia may serve as a definition for a sidereal, or Copernican, reference frame whose origin would be the center of mass of the solar system and whose three axes would be determined by three fixed stars in our galaxy.

With respect to such a reference frame, a free particle—meaning one subjected to $\mathbf{F} = 0$—has zero acceleration ($\mathbf{a} = 0$).

Newton's Second Law

In its original form this law states that the rate of change of motion of a body is proportional to the force that acts on it and takes place in the direction of this force.

This law has also met with widespread acceptance today: "The total force acting upon a particle is equal to the product of its mass and its acceleration": $\mathbf{F} = \mathbf{ma}$.

Stated somewhat differently (since the mass m of the particle is constant): ''The total force acting upon a particle is equal to the time derivative of the said particle's momentum (m$\mathbf{v}$ = $\mathbf{p}$)'':

$$\mathbf{F} = d\mathbf{p}/dt$$

Many scientists use this last equation as the sole definition of force.

Newton's Third Law, or Law of Equal Pairs of Forces

Newton's original statement reads as follows: For every action there is always an equal and opposite reaction; or stated differently: The mutual forces exerted by two bodies on each other are always equal, but occur in opposite directions.

This law is interpreted today as: ''When two particles (1) and (2) act upon one another, the force $\mathbf{F}_{12}$ exerted by particle (1) on particle (2) is equal to, but acts in a direction opposite to the force $\mathbf{F}_{21}$ exerted by the second particle (2) on the first (1)'': $\mathbf{F}_{12} = -\mathbf{F}_{21}$.

A number of comments are in order here. To begin with, concerning the first law, prior to Newton, Galileo had said that a body subjected to a zero resultant force should have constant velocity. This statement, also called the law of inertia, defines a special category of reference frames or Galilean reference systems. These are any reference bodies moving at constant speed (in direction and modulus) with respect to the sidereal, or Copernican, reference frame defined previously.

Second, as regards Newton's second law, and in light of what has just been said, we must note that this law has meaning only in a Galilean reference frame and when applied to a particle of mass m and expressed in the equation $\mathbf{F}$ = m$\mathbf{a}$ or $\mathbf{F}$ = d$\mathbf{p}$/dt. Referring to chapter 3, which deals with inertial and Coriolis forces, we note that if we take into account these inertial forces, then the basic law of dynamics as applied to a particle of mass m becomes:

$$\mathbf{F} + \mathbf{F}_0 = m\mathbf{a}_r,$$

where $\mathbf{F}_0$ represents the sum of inertial and Coriolis forces and $\mathbf{a}_r$ is the relative acceleration with respect to a non-Galilean reference body undergoing acceleration. In fact, we do get the same basic law, since:

$$m\mathbf{a} = m\mathbf{a}_r - \mathbf{F}_0,$$

where $\mathbf{a}$ is the acceleration of the particle with respect to a Galilean reference system.

Third, as with the first and second laws, modern interpretation of the third law requires that *bodies* (as originally stated by Newton) should be particles or point masses. Hence with respect to the law of equal action and reaction, if the forces $\mathbf{F}_{12}$ and $\mathbf{F}_{21}$ must be equal and opposite they must also act along the same line. From these considerations we have the implicit assumption that Newtonian forces of interaction are central forces. Any other forces of interaction, such as the magnetic force acting upon a charged particle moving within a magnetic field $\mathbf{B}$ or the Coriolis force acting on a marble moving in free fall toward the surface of the earth, are certainly forces, but non-Newtonian ones. On the other hand, the universal force of attraction between any two point masses separated by a distance r ($F = Gm_1m_2 / r^2$) or the Coulomb force of attraction or repulsion between two charges q_1 and q_2 separated by a distance r ($F = q_1q_2/4\pi\epsilon_0 r^2$) meet the two requirements stipulating that (1) action and reaction be always equal and (2) the line of action of both forces be unique (central forces).

We should note, however, that Newton's original statements do not require that the forces of interaction between two bodies act along the same line of action. We shall see in chapter 3 ("Inertial Forces") and chapter 4 ("Newton's Third Law" [Generalized]) that there are, in fact, many instances in nature where forces of interaction between two bodies are not central. And yet it will be seen that these forces respect the three laws of motion, even if two of these laws—the first and the third—do require some updating in their formulation.

The Origin of Forces

Up to the present we have been considering a few examples of forces such as muscular thrust or traction, the forces of gravitational interaction between two bodies, or electromagnetic forces. In many texts, force is defined starting with Newton's second law (see the preceding section), meaning that which causes the change in a body's momentum. This definition is correct, but not sufficient. All forces must obey the three laws of motion. Everything surrounding us is in constant change: Flowers bloom, planets follow their prescribed orbits, and bombs explode. Such changes can be explained only by reference to motion, and it is entirely understandable that one might want to explain such observable phenomena by the forces causing them.

At the present time, all the forces of interaction observed between macroscopic or microscopic objects are thought to be derived from one or another of the five basic forces. These are: (1) the force of gravity, (2) the inertial and Coriolis forces, (3) the electromagnetic forces, (4) the weak nuclear force, and (5) the strong nuclear force.

Gravity

The force of gravitational attraction exerted by a particle with a mass m_1 upon another particle with a mass m_2 is given by:

$$\mathbf{F}_{12} = -\frac{G\,m_1 m_2}{r^2}\,\frac{\mathbf{r}_{12}}{r}$$

where $\mathbf{r}_{12} = \mathbf{A}_1\mathbf{A}_2$ is the vector position of particle (2) with

5

respect to particle (1), r is the distance of separation of the particles and G is the constant of universal attraction. It is a central force, and the same formula may be used as long as the masses m_1 and m_2 are spherical and uniformly distributed. We shall see further on that when the form of an interacting body is no longer strictly spherical—as in the case of the Earth, which is an ellipsoid flattened at the poles and subjected to the sun's attraction—gravitational interaction may result in a noncentral force of interaction and a couple. See the examples of the precession of the equinoxes in chapter 4 and of the secular slowing down of the Earth's rotation in chapter 6.

In modern textbooks inertial forces are not included as basic forces insofar as these forces are excluded from the basic law of dynamics if a Galilean reference frame is involved. But if a particle moves with respect to a rigid body to which any type of motion has been imparted (for instance, rotation and translation), there should be forces of interaction arising from the accelerated motions of the rigid body and/or the relative motion of the particle with respect to the body. We shall see in chapter 3 that these inertial forces are equivalent or similar to gravitational forces and that, like gravitational forces, they obey the generalized law of equal action and reaction.

Electromagnetism

Electromagnetic forces are composed of two constituents: the Coulomb force and the magnetic force. Let there be two charge points q_1 and q_2. The Coulomb force of attraction or repulsion is given by the formula:

$$\mathbf{F}_{12} = \frac{1}{4\pi\epsilon_0} \frac{q_1 q_2}{r^2} \frac{\mathbf{r}_{12}}{r} ,$$

where ϵ_0 is the basic electromagnetic constant and the other symbols have been defined previously. Attraction occurs when the two charges have opposite signs; repulsion occurs when the two charges have the same sign.

The magnetic force $\mathbf{F}_m$, which acts upon a particle moving at a velocity $\mathbf{v}$ with respect to a reference frame R in which there exists a magnetic field $\mathbf{B}$, is given by the vector product:

$$\mathbf{F}_m = \mathbf{q}\,\mathbf{v} * \mathbf{B}$$

Unlike the Coulomb force, the magnetic force is non-central. We shall demonstrate in chapter 4 that this magnetic force nevertheless respects the generalized law of equal action and reaction.

The Strong and Weak Nuclear Forces

These forces are not easily measured. The forces acting among the particles contained in the nucleus of an atom belong to the strong nuclear forces, and much remains unknown about them. They are about a hundred times stronger than electromagnetic forces, but are active over vastly reduced distances: roughly $1/100,000$ the diameter of an atom. This force is responsible for the mutual attraction between tightly bound nucleons (protons and neutrons), but disappears whenever nucleons are separated. It is thus this force that holds the nucleus of an atom together.

The weak nuclear force, on the other hand, is also active over small distances, but is extremely weak: about one millionth of the strength of the strong nuclear force. The weak nuclear force is responsible for the transformation of a neutron into a proton in the reaction:

$$n \longrightarrow p^+ + e^- + \bar{\nu}$$

A neutron (n) is transformed into a proton (p^+), and in the process an electron (e^-) and an antineutrino ($\bar{\nu}$) are also released. Neutrinos and antineutrinos are particles which can be acted upon only by the weak nuclear force. Recent discoveries, leading to a Nobel prize in 1984, have shown that weak nuclear interaction forces are conveyed by photons.

We shall limit our discussion of such forces to these considerations inasmuch as the main area of interest in this book lies in the forces of gravity, electromagnetism, and inertia.

Electromagnetic forces are much stronger than gravitational forces, the ratio between their strengths being on the order of 10^{40}. When we push a book across a table, we exert an electromagnetic force on the book. The atoms in our fingers repel the atoms in the book. In the final analysis, all forces exerted by man are electromagnetic in nature. However, if the forces of gravity seem to us to be so great, it is due to the Earth's considerable mass. Gravitational forces result from the reciprocal action of two masses, but unless one of these masses is huge, the interaction force is usually so weak that we do not notice it.

Laws of Conservation of Linear Momentum, of Angular Momentum, and of Energy in an Isolated System

There is a law called Galilean relativity according to which all basic laws of mechanics have the same form, regardless of the Galilean reference frame used to express them. Of all these laws, it is the laws of conservation of linear momentum, of angular momentum and of energy in an isolated system that are particularly important. In fact, these laws apply to all areas of classical, relativistic, and quantum physics and are more or less connected with the so-called invariance laws through translation or rotation in space, or through translation in time.

Conservation of Energy

According to the law of conservation of energy, the total energy in the universe remains constant and independent of time. This law of conservation of energy in an isolated system may be stated as follows: "For a system of particles whose interactions do not explicitly depend on time, the total energy of the system will be constant. This energy is made up of kinetic energy, potential energy (gravity, electromagnetic . . .), heat energy, etc. and is measured either in a Galilean reference frame or in the frame of the *source* . . ."

At first glance, the law of conservation of energy appears to be only an integrated form of Newton's Second Law $\mathbf{F} = \mathbf{ma}$, enabling us to describe various motions without reference to time. To illustrate this property, let us consider a simple isolated system composed of a point mass subjected to the elastic attraction of a massless spring. The total energy of this system as measured in a Galilean reference frame is constant. At any instant we shall have:

$$\mathbf{E}_{total} = \tfrac{1}{2} m v^2 + \tfrac{1}{2} k x^2 = \text{constant}$$

Where $E_c = \tfrac{1}{2} m v^2$ is the kinetic energy of the point mass m and the following term is the potential energy of elastic interaction of mass m. Taking the time derivative of the above relation, we obtain:

$$mv \, (dv/dt) + kx \, (dx/dt) = 0,$$

and as $v = dx/dt$, after simplifying with v, we have:

$$m \, (dv/dt) = -kx,$$

which is really the expression of the basic law of dynamics applied to mass m ($-kx$ = the force exerted by the spring on mass m).

This result may be generalized so as to encompass an isolated system made up of n particles. The law of conservation of energy may also be said to be linked to the law of invariance through translation in time.

Conservation of Linear Momentum

In his *Principes de philosophie*, Descartes wrote: "Although motion is but one aspect of matter in movement, it nonetheless does possess a certain quantity . . . which neither increases nor decreases . . . regardless of there being now more or less of this quantity in various parts. It is for this reason that, when one part of matter moves twice as fast as another, this other part being twice as large as the first, we are led to think that there is just as much motion in the smaller as there is in the larger part, and furthermore, that when motion decreases in one part, the motion in the other part increases proportionally.''

Here we have an admirable statement of the law of conservation of momentum—even if Descartes had not perceived that momentum p of a mass point m is a vector quantity today written as $\mathbf{p} = m\mathbf{v}$, where $\mathbf{v}$ is the velocity of m with respect to a Galilean reference frame.

Applied to the collision of two bodies (1) and (2), this law of conservation of momentum states that the sum of momenta *after* collision is equal to the sum of momenta *before* the collision:

$\mathbf{p}_1$ (before) $+ \mathbf{p}_2$ (before) $= \mathbf{p}_1$ (after) $+ \mathbf{p}_2$ (after)

The law of conservation of momentum is deduced from Newton's second and third laws. In fact, starting with the basic law of dynamics and for a finite interval of time Δt, we can state (the system of [1] and [2] being isolated):

$$\mathbf{F}_{21} = m_1 \frac{\Delta \mathbf{v}_1}{\Delta t} \text{ and } \mathbf{F}_{12} = m_2 \frac{\Delta \mathbf{v}_2}{\Delta t}$$

According to Newton's third law, the force $\mathbf{F}_{12}$ exerted by body (1) upon body (2) is equal to and opposite $\mathbf{F}_{21}$, i.e.: $\mathbf{F}_{12} + \mathbf{F}_{21} = 0$, hence:

$m_1 \Delta \mathbf{v}_1 + m_2 \Delta \mathbf{v}_2 = 0$; integrating, we obtain:

$(m_1 \mathbf{v}_1 + m_2 \mathbf{v}_2)_{\text{before}} = (m_1 \mathbf{v}_1 + m_2 \mathbf{v}_2)_{\text{after}}$

At each instant, the sum of momenta $(\mathbf{p}_1 + \mathbf{p}_2)$ of the two particles is conserved. This result may be generalized and applied to an isolated system containing n particles.

The vector sum of momenta in an isolated system of n particles (as measured in a Galilean reference frame) is conserved over time.

10

The law of conservation of momentum is linked to the law of invariance through translation in space.

Conservation of Angular Momentum

One of the consequences of the theorem of angular momentum as applied to a material system is that total angular momentum of an isolated system remains constant over time. This property is expressed in classical mechanics with the restrictive hypothesis that the forces of interaction involved must be central forces. We shall see in chapter 4 that the law of conservation of angular momentum in an isolated system results from Newton's generalized third law. In fact, all interactions in nature occur through exchanges of linear momentum (forces) or through exchange of angular momentum (couples or moments of forces). This means that in the generalized statement of the law of equal action and reaction we must reckon with interactions between bodies occurring by means of couples or moments of force.

Newton's laws of motion are postulates. Similarly, the law of conservation of total angular momentum in an isolated system (as measured relative to a Galilean reference frame) is a law *founded on experimental evidence* and not on mathematical demonstration. To date, no experiment has yet cast doubt on the law of conservation of angular momentum. What is more, by bringing in the law of invariance through rotation in space, this law may be extended to the whole of the universe. In nuclear physics, however, whenever discrepancies are found in energy tallies or angular momentum tallies, one usually prefers to "invent" new particles rather than admit that these laws of conservation have been broken.

Such was the case when the Swiss physicist Pauli, to explain the transformation of a proton into a neutron:

$$p^+ + e^- \longrightarrow n,$$

and the apparent violation of the law of conservation of angular momentum, came up with the idea of writing the reaction differently:

$$p^+ + e^- \longrightarrow n + \nu$$

The law could now be respected: ν was a new particle baptized "neutrino" having zero mass, zero electric charge, a spin equal to $\frac{1}{2}$, and a leptonic number equal to $+1$. Subsequent investigations led to proof of the new particle's existence.

Interactions Involving Torques or Moments of Forces

In everyday life there are countless instances in which several thrust or traction forces act upon an object, even if their vector sum turns out to be zero. You exert a couple on a corkscrew when you drive it down through the cork in a bottle of wine.

A couple is defined as the combination of two forces, equal in magnitude but opposite in direction, acting along two parallel straight lines. The resultant, or vector sum of the two forces, is obviously zero ($\mathbf{F}_1 + \mathbf{F}_2 = 0$), and will therefore have no effect on the translation motion of the rigid body. However, this couple tends to cause the object to rotate around an axis perpendicular to the plane defined by the forces $\mathbf{F}_1$ and $\mathbf{F}_2$. If we designate the application points of these forces respectively by A_1 and A_2 belonging to the rigid body under consideration ($\mathbf{r}_1 = \mathbf{O}\,A_1$ and $\mathbf{r}_2 = \mathbf{O}\,A_2$), then at a point O in space the couple is described as the sum of the moments of the two forces given by the vector products:

$$M_0 = \mathbf{r}_1 * \mathbf{F}_1 + \mathbf{r}_2 * \mathbf{F}_2 = (\mathbf{r}_1 - \mathbf{r}_2) * \mathbf{F}_1 = A_2 A_1 * \mathbf{F}_1.$$

Inasmuch as the couple $\Gamma = A_2 A_1 * \mathbf{F}_1$ is independent of the position of the chosen point O, it may be measured at any point in space and will have as an axis a straight line perpendicular to the plane of the two forces $\mathbf{F}_1$ and $\mathbf{F}_2$. Its modulus will be the product of the distance between the two supports and the modulus of one of the forces (F_1).

The use and interest of the moment of a force ($\mathbf{r_1}*\mathbf{F_1}$) at point O may be illustrated by means of an example. When we open (or close) a door, we exert a thrust (or traction) force on the handle in such a way that the door swings on its hinges. The effect obtained is a change in the angular momentum of the door in its rotation motion. This change takes place thanks to the application of the moment of the thrust (or traction) force with respect to the vertical axis of rotation. For any given moment, the longer the lever arm—i.e., the distance between the axis, passing through the hinges, and the handle—the less will be the effort required.

In practice, it often happens that interactions between bodies occur by means of couples. Such is the case for a motor composed of a *stator* exerting a couple Γ on a *rotor*. The rotor also exerts a couple $-\Gamma$ on the stator. Similarly, when you hold a small electric motor in your hand—an electric coffee grinder, for instance—you distinctly feel this couple ($-\Gamma$) being exerted on your hand as soon as the motor is switched on.

And likewise, the cabin of a helicopter exerts on the rotor a couple (Γ) whose vertical axis passes through G (the center of mass). The cabin, in return, is subjected to a couple ($-\Gamma$), which tends to cause the rotor to turn in the opposite direction. In order to keep the helicopter in its line of flight, the reaction couple ($-\Gamma$) must be canceled out. This is achieved by means of a small propeller at the rear of the helicopter at point P. The propeller, whose horizontal axis is perpendicular to the line of flight, sets up a horizontal force $\mathbf{f}$ whose moment with respect to the axis of the rotor cancels out the couple ($-\Gamma$):

$$-\Gamma + \mathbf{GP} * f = 0,$$

where the vector $\mathbf{GP}$ is horizontal and parallel to the line of flight.

A record player turntable, a turnstile, a top, and/or gyroscope wheel are all set into rotary motion on their axes by applying a couple (from outside). As in the previous cases, the

13

law of equal action and reaction forces applies to couples.

In chapters 3 and 4 we shall see further important instances where interactions between physical bodies occur by means of noncentral forces and couples. This is the case for interactions arising from the Coriolis force (see chapter 3) exerted on a marble in free fall by the Earth rotating on its axis (one complete rotation every twenty-four hours). The same holds true for the electrostatic interaction between a charge point and an electric dipole, as well as for electric and magnetic interactions.

And finally, the same principle applies to Earth–sun and Earth–moon gravitational interactions. Indeed, chapter 4 will show that the sun exerts on the Earth not only a gravitational force of attraction—one of whose components is noncentral—but also a couple whose axis is perpendicular to the plane defined by the two centers and by the axis of the Earth's poles. It is this couple, together with the couple exerted by the moon on the Earth for similar reasons that cause the retrograde motion observed in precession of the equinoxes.

In this section we have first examined the simple cases of interactions arising from couples. We have noted that the law of equal action and reaction forces in these simple instances may be easily generalized: $\Gamma_{12} + \Gamma_{21} = 0$. The examples that we cited at the end of the section—and which will be studied in more detail in the following chapters—are somewhat more complex in that they involve noncentral interaction forces. Generalizing Newton's third law will prove to be slightly more delicate and will be taken up in the section that follows. We shall also deal with generalizations of Newton's other two laws.

Newton's Three Laws of Motion (Generalized)

In keeping with the observations made throughout this first chapter, we shall now attempt to generalize Newton's three laws of motion without distorting the original meaning, keeping them in perfect harmony with the laws of conservation of linear momentum, angular momentum, and energy.

Newton's First Law (Generalized)

In the absence of any outside force, a body remains in a state of rest or in constant motion. This same body will remain in a state of uniform rotary motion unless a couple produces a change in this state.

The first part of the statement repeats Newton's words; it corresponds to the law of inertia and accounts for the translation motion of a rigid body subjected to zero resultant force:

$$\mathbf{a}_G = 0 \qquad \text{if} \qquad \Sigma \mathbf{F}_{\text{ext.}} = 0.$$

where $\mathbf{a}_G$ is the acceleration of the center of mass G of the body, relative to either the Copernican reference frame defined above or any other reference body moving at constant velocity with respect to the former Galilean reference body.

The second half of the statement involves possible rotary motion of the rigid body—as in the case of spherical bodies such as the planets and stars and symmetrical tops:

$$I_\Delta \omega = \text{constant if } M_G (\mathbf{F}_{\text{ext.}}) = 0,$$

where I_Δ is the moment of inertia of the object relative to its rotation axis, ω is the angular rotation vector of the object relative to this axis, and $M_G (\mathbf{F}_{\text{ext.}})$ is the moment at G of the external forces acting on the object.

The Earth itself would, at first glance, appear to continue in its motion of constant rotation on its polar axis, making one complete rotation every twenty-four hours, and with no couple intervening to produce a change. As a result, it would seem, the axis of the Earth's poles should have a fixed direction relative to the Copernican reference frame. In fact, however, this very axis undergoes an extremely slow motion of precession not unlike that experienced by a spinning top (see chapter 4: "Precession of the Equinoxes"). This precession is due to very weak couples exerted by the sun and the moon on the Earth.

The Second Law of Motion Applied to Any Material System

Newton's original statement is not precise enough to be easily generalized. When he wrote it, Newton may very well have been thinking of "body" as being an object whose mass is constant.

The statement of the basic law of dynamics may be generalized here so as to apply to any material system whatever (total mass $M = \Sigma \, m_i$, either constant or varying in time):

The sum of the external forces acting on any material system (total mass M) is equal to the sum of the products $m_i \, a_i$ (m_i = elementary mass of one component in the system, a_i = acceleration of m_i relative to a Galilean reference body):

$$\Sigma F_{ext.} = \Sigma \, m_i \, a_i \qquad (1)$$

For: $\Sigma \, F_{int.} = 0$, where $F_{int.}$ represents the internal forces interacting within the material system.

The equation (1) represents the theorem of the center of mass when applied to a material system, the total mass of which remains constant. Indeed, in this case: $\Sigma \, m_i a_i = dP \, / \, dt$, where $P = \Sigma \, m_i v_i = M \, v_G$ is the total linear momentum of the material system.

The next chapter will show that the same equation (1) may be applied to a material system whose mass varies over time. However, the theorem of the center of mass is no longer involved in this instance, for:

$$\Sigma \, m_i \, a_i \neq dP \, / \, dt$$

(Cf. the example of the Atwood machine and the conveyor belt.)

The theorem of angular momentum applied to a material system at a point G (= center of mass) can be easily deduced

from the foregoing equation, yielding:
$$M_G \ (F_{ext.}) = dL_G \ / \ dt,$$
where $L_G = \Sigma \ G \ A_i^* \ m_i v_i$ is the angular momentum of the system measured at point G; A_i is the point at which m_i is found; v_i is the velocity of m_i relative to a Galilean reference frame, and M_G $(F_{ext.})$ is the moment at point G of the external forces acting upon the system.

Using Newton's second (generalized) law, we note that we can find the classic expressions involving the theorems of the center of mass and angular momentum as applied to a material system with constant total mass. Whenever a solid body is involved, these two vector equations make it possible to determine the equations for its translational and rotational motions.

Newton's Third Law (Generalized)

As we shall see in detail in chapter 4, it is especially this law of equal action and reaction forces which has come under attack in recent textbooks. Generalization of this law is thus necessary if we wish to avoid contradiction with the laws of conservation of linear momentum, angular momentum, and energy in an isolated system. Let us now look at the generalized statement:

The bodies act upon one another in pairs of actions that are always equal in magnitude and opposite in direction. The action of object A on object B is revealed by the presence of either a force F_{AB} or a couple Γ_{AB} or a combination of force F_{AB} and couple Γ_{AB}.

Mathematically, the generalized law of equal action and reaction can be expressed by two vector equations:
$$F_{AB} + F_{BA} = 0 \qquad \text{and} \qquad (M_{AB})_0 = (M_{BA})_0,$$
where the first equation is the classic expression of Newton's third law (but extended to noncentral interaction forces) and where the

second equation is an equivalence extended to moments of mutual actions and couples measured at a point 0 in space.

The first part of the statement is Newton's, while the second part is intended to account for possibilities of interaction between two bodies through couples or exchange of angular momentum.

The study of the generalized law of equal action and reaction is so important that we devote an entire chapter to it (chapter 4). From this study there emerges, in fact, a definite analogy in the ways the three basic types of forces—gravity, electromagnetic, and inertia—behave.

Chapter 2
Variable Mass Systems

The vast majority of material systems that we meet in physics are systems that have constant total mass. However, we can choose the system arbitrarily, including or excluding any particle as we see fit. But if forces are applied to our system by particles entering or leaving the system or if other external forces come to act upon the system, we must bring these two types of forces into the basic law of dynamics.

It must be remembered that in a Galilean reference frame, at any given instant t, the expression of Newton's second law is provided by the formula:

$$\Sigma \, m_i \, (dv_i/dt) = M \, (dv_G/dt) = dP/dt = \Sigma \, \mathbf{F}_{ext.} \qquad (1)$$

where m_i is the mass of the ith particle; $M = \Sigma \, m_i$ is the constant total mass of the system; $v_i \; v_G$ are the respective velocities of mass m_i and of the center of mass G relative to a Galilean reference frame.

$\mathbf{P} = \Sigma \, m_i v_i = M \, v_G$ is the total linear momentum on the system, and $\Sigma \, \mathbf{F}_{ext.}$ is the sum of the external forces acting on the system and whose application point is the center of mass G.

Equation (1) expresses the basic law of dynamics as it applies to a material system composed of n particles, the sum of whose masses $M = \Sigma \, m_i$ remains constant over time.

The external forces $\Sigma \, \mathbf{F}_{ext.}$ that act on the components of the material system obey the law of equal action and reaction. In this chapter we shall use the law only in its simplified form: (1) the force exerted by a particle n on a particle k is equal to but opposite that exerted by k on n; (2) the forces of interaction are assumed to be central forces, or, in the event of contact forces, the action and reaction supports are borne by the same straight line.

The Basic Law of Dynamics as Applied to a Variable-Mass System

There is some confusion in the scientific literature about what should be the correct equation to apply in the case of motion involving variable-mass systems, meaning systems that may either gain or lose particles. Certain writers assert, for instance, that Newton's second law can only be applied to systems with constant total mass. Others maintain that the law applies to any solid system (with mass M being either variable or constant) in the form: $d(\Sigma\ m_i v_i)/dt = \Sigma\ F_{ext}$. They go on to claim that what is involved is a generalization of the postulate as formulated by Newton, with the rate of change in a body's motion being proportional to the force acting upon it and in the direction of this force.

To reconcile these two otherwise conflicting viewpoints, I would like to show in this chapter that, using several typical examples, Newton's second law may apply to any material system with total mass being either constant or variable and that its expression at any given instant t in a Galilean reference frame will be:

$$\Sigma\ m_i\ (dv_i/dt) = \Sigma\ F_{ext.} = F_{interactions} + F_{others} \qquad (2)$$

where $F_{interactions}$ represents the external forces acting on the particles of the system (with total mass being variable either by addition or subtraction of particles) by those particles that enter or leave the system at time t, and where F_{others} represent the other external forces acting on the particles in the system different from the former ones.

For variable mass systems, the definition of the center of mass has hardly any use; equation (2) does not represent the theorem of the center of mass, as was the case for equation (1), pertaining to systems with constant total mass. In fact, when systems having total variable mass are involved, the time derivative of total linear momentum ($\mathbf{P} = \Sigma\ m_i\mathbf{v}_i$) is not equal to the term $\Sigma\ m_i\ (d\mathbf{v}_i/dt)$: $\Sigma\ m_i\ (d\mathbf{v}_i/dt) \neq d\mathbf{P}/dt$.

Finally, the term $\mathbf{F}_{\text{interactions}}$ is found by applying Newton's third law to two interacting objects: the mass M(t) of the system and the increment in mass Δm entering or leaving the system during an interval of time Δt. This method that I propose for determining the external forces is not original, insofar as Newton himself had already set forth the principle in his book *Philosophiae Naturalis Principia Mathematica*. The discoveries of the law of universal attraction—i.e., between any two bodies there exists a force of attraction that is in proportion to the product of the masses of the objects and in inverse proportion to the square of the distance between them—and the law of equal action and reaction allowed Newton to go beyond the mere notion of a two-body system. He was now able to show that the Earth, in its simple movement around the sun, followed an elliptical orbit, one of whose foci was the center of the sun (second law, equation [1]). He further demonstrated that the force exerted by the sun on the Earth was due to the law of universal gravitation and, finally, that if the sun attracted the Earth, the Earth also attracted the sun with equal and opposite force.

We should note that in analytical mechanics it is customary to call inertia force the term $(-m_i\mathbf{a}_i)$ where $\mathbf{a}_i = d\mathbf{v}_i/dt$ is the acceleration of particle m_i measured in a Galilean reference frame. Thus the basic law of dynamics applied to a particle can be written: $(-m_i\mathbf{a}_i) + \mathbf{f}_i = 0$, i.e.:

absolute force of inertia + force acting on $m_i = 0$.

Seen in this way, the two equations (1) and (2) may be written as vector sums of the preceding equation:

Σ absolute forces of inertia $+ \Sigma$ external forces $= 0$.

However, the analogy is not complete. As we have said above, equation (1) applied to a material system with constant total mass represents the center of mass theorem. This is not the case with equation (2) when applied to a variable mass system. What is more, Lagrange's equations cannot be applied to a system with variable total mass.

And finally, let us consider particular cases involving material systems whose components are motionless with respect to each other. The change over time in mass m(t) of such a system stems from the fact that at a given instant t certain particles enter the system or leave it. Such addition or subtraction may occur continuously (as in the case of a rocket) or all at once (for instance, when ballast is jettisoned from an air balloon). Newton's second law as applied to such a system in a Galilean reference frame and at a given instant t in time will be written:

$$m(t) \, \frac{d\mathbf{v}}{dt} = \mathbf{F}_{\text{interactions}} + \mathbf{F}_{\text{others}}, \qquad (3)$$

where $\mathbf{v}$ is the velocity common to the various components in the system whose mass is m(t) and where the forces are defined as in equation (2) above. Equation (3) may be used in order to study translational motion in bodies whose mass increases or decreases in time, as we shall see in the following examples.

The Problem of the Rocket

Of the many variable mass systems, the one that most challenges our imaginations today is certainly the rocket used to convey spacecraft. I take care in saying "today" in light of what one great physicist, J. C. Maxwell, wrote in a letter to fellow physicist P. G. Tain dated February 1878: "I do not know how to apply the laws of motion to objects whose mass undergoes change, inasmuch as no experiments involving such bodies have ever been conducted, just as no experiments have ever been carried out on objects whose mass would be negative.'' The first experiments conducted with V2 rockets used by the Germans to bomb England took place during the Second World War, in 1943. The principle of jet propulsion is straightforward. When gases are burned in a combustion chamber, they acquire high speeds from thermal agitation. They are then expelled through the exhaust nozzle. It is the high speed with which the exhaust gas leaves the rear of the engine that results in the force that pushes the aircraft forward. This force is called the forward thrust ($\mathbf{F}_{\text{interactions}}$) in equation (3).

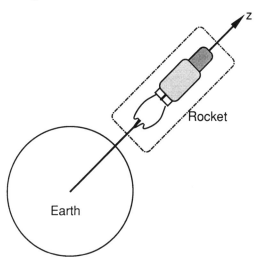

Figure 1. The Rocket

Determining the Thrust Force

According to the law of equal action and reaction forces, the thrust force is equal to but opposite that force exerted by the

23

rocket on the exhaust gases. The two interacting bodies are the rocket with a mass $m(t)$ at a given instant t and the increment of mass $(-dm)$ of the exhaust gases during the interval of time dt (see figure 1). The quantity $(-dm)$ is positive because the rocket's mass decreases by an amount dm during the same interval of time dt.

Let us imagine that these two objects are isolated and that the rocket's trajectory is rectilinear (axis Oz). In other words, no outside force acts on either the rocket or the exhaust gases. The momentum of the exhaust gases, with respect to the Oz axis but in a reference frame connected with the rocket at time t, is $-\mathbf{u}dm$, where $\mathbf{u}$ is the velocity of the exhaust gases relative to the rocket. The momentum of the rocket was zero at time t—as defined by the reference body chosen and no longer linked to the rocket at time $t + dt$—but now increases, at first by an amount $m(t) \, d\mathbf{w}$ during the interval of time dt. Since the interaction takes place through an exchange of momenta, the sum of the two preceding quantities must remain zero during the interval of time dt:

$$m(t) \, d\mathbf{w} - \mathbf{u} \, dm = 0$$

After dividing by dt and tending dt toward zero, because the rocket's velocity increases by $d\mathbf{w}$, the rocket at a given time t is subjected to a force $\mathbf{F}_{interactions}$ exerted by the exhaust gases:

$$\mathbf{F}_{interactions} = \mathbf{u} \, (dm/dt)$$

This thrust force acts in a direction opposite that of the velocity $\mathbf{u}$ of the exhaust gases (with respect to the rocket).

In reference to the foregoing demonstration, much has been written about applying the law of conservation of momentum. Here it is necessary to insist on three main points:

First, that the equation: $m(t) \, d\mathbf{w} - \mathbf{u} \, dm = 0$ is correct even in a non-Galilean reference frame linked to the rocket at time t. This is not usually the case for the law of conservation of momentum in an isolated system.

Second, the equation written in the form $\mathbf{F}_{interactions} = \mathbf{u}$

(dm/dt) remains correct even if the two-body system, such as rocket + exhaust gases, is not isolated. Since both rocket and exhaust gases are subjected to the gravitational attraction of the Earth, the thrust force $\mathbf{F}_{\text{interactions}}$ remains equal to $\mathbf{u}$ (dm/dt) thanks to Newton's third law.

Third, interactions between the two bodies—rocket and exhaust gases—takes place through the mutual exchange of momenta by means of contact forces.

Equation Governing the Rocket's Motion

In order to determine the equation for the rocket's motion, we must take into consideration the system composed of the rocket having variable mass m(t) and then choose a reference body, preferably Galilean, that will be linked to the vertical departure axis Oz and whose direction will remain fixed in a Copernical reference frame following the rocket's lift-off (see figure 1). With respect to this reference frame, the basic law of dynamics (equation [3]) applied to the rocket at time t will be written:
$$m(t)\ (dv/dt)\ =\ \mathbf{u}(dm/dt)\ +\ m(t)\ \mathbf{g}\ +\ \mathbf{R}$$
where $\mathbf{g}$ is the Earth's field of gravitational acceleration and $\mathbf{R}$ is a force arising from air resistance for such time that air exists.

We observe that in this equation all the forces acting on the rocket—thrust, the earth's gravity, air resistance—are external forces and that m(t) (dv/dt) is the rocket's absolute force of inertia with a change in the sign.

Atwood's Machine

Atwood's machine is illustrated in figure 2. Many texts include a detailed description of simplified models that make it possible to investigate motion in this machine. Such models customarily rely on several assumptions. First: the cord of length L (1) is inextensible, (2) has negligible mass, and (3) experiences no slippage as it moves over the pulley. Second, it is assumed that the pulley turns on its axis without fricton and (more often than not) has negligible moment of inertia. These assumptions enable us to simplify the solution of the problem created by Atwood's machine.

However, the discussion of a model in which the cord has a nonnegligible mass $m = \lambda L$ where λ is equal to mass per unit of length, does not exist in the scientific literature. But the very such discussion is highly relevant since it affords the scientific proof that Newton's second law of motion as applied to a variable mass system should be written in the form of equation (3) above and not in the form $d(m\mathbf{v})/dt = \Sigma \mathbf{F}_{ext}$.

Suppose that motion occurs in the direction shown by the arrow in figure 2, making mass m_1 fall toward the ground while mass m_2 is pulled upward. Now consider two variable mass systems, the first composed of mass m_1 and the cord up to point A (total mass $m_1 + \lambda z$) and the second made up of mass m_2 and the second part of the cord starting from m_2 and extending to point B (total mass $m_2 + \lambda[L - z]$). The mass of the cord along arc AB of the pulley is negligible, as is the pulley's moment of inertia. The force exerted by system 2 on system 1 at A (T_{21}) is equal to and opposite the force exerted by system 1 on system 2 at B (T_{12}) in accordance with the law of equal action and reaction applied for the determination of $\mathbf{F}_{interactions}$ in equation 3. As for the term $\mathbf{F}_{others}$ in the same equation, what is involved is total weight, meaning $(m_1 + \lambda z)g$ for system 1 and $(m_2 + \lambda[L - z])g$ for system 2.

Applying the basic law of dynamics (equation [3]) to the two variable mass systems 1 and 2 gives us, respectively:

$$(m_1 + \lambda z)a = -T_{21} + m_1 g + \lambda z g$$
$$- (m_2 + \lambda(L - z)) a = -T_{12} + m_2 g + \lambda (L - z) g$$

From these two equations, it is possible to derive the expressions of the acceleration a of the machine as well as the tension

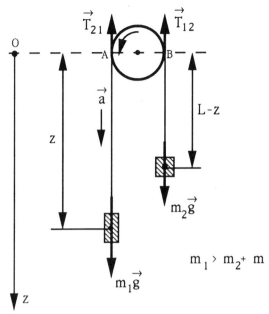

Figure 2. Atwood's Machine

T_{12} ($= T_{21}$) of the cord at the level of the pulley, giving:

$$a = \frac{m_1 - m_2 + \lambda(2z - L)}{M} \quad \text{with: } M = m_1 + m_2 + \lambda L,$$

$$2T_{12} = \frac{M^2 - (m_1 + \lambda(2z - L) - m_2)^2}{M} \ g$$

If the mass per unit of length of the cord ($\lambda = 0$) is negligible, we find that we have the classic results. On the other hand, if the mass of the cord cannot be disregarded considering m_1 and m_2, we note that the acceleration a and the tension T_{12} will vary, albeit weakly, with the height z of mass m_1.

Proof of the Equation m(t) dv/dt = Σ F_e

The same problem can, of course, be solved by applying the basic law of dynamics to systems 1 and 2 as previously defined, but by writing (incorrectly): $d(mv)/dt = \Sigma\, F_e$.

From the two equations of motion, one can derive:

$$a = \frac{m_1 - m_2 + \lambda(2z - L)}{M}$$

$$2T_{12} = \frac{M^2 - (m_1 + \lambda(2z - L) - m_2)^2}{M}\, g - 2\lambda v^2$$

Although the expression for the acceleration a is identical to that in the previous section, the result as far as the tension T_{12} goes is physically inadmissible insofar as the tension on the cord can never be a function of the velocity v of masses m_1 and m_2. In addition, if the preceding formula were correct, we could select respective values for m_1, m_2 and λL in such a way that the tension T_{12} would be canceled out upon completion of the motion, and this would be absurd. Since the result for the tension T_{12} is not acceptable, this means that our initial theory $(d[mv]/dt = \Sigma\, F_{ext.})$ is incorrect. Naturally, if it were possible to measure the tension of the cord on the pulley at each instant, the previous conclusions could be definitively confirmed—or invalidated.

Encke's Comet

A recent article (1) brought to light a variable mass system in celestial mechanics: Encke's Comet. In his article, F. L. Whipple shows that the variations observed in this comet's period of rotation as it revolves around the sun can be attributed to the rotation of its ice nucleus and the thrust caused by gases evaporating on its surface (a phenomenon called sublimation).

The comet was discovered by Pierre Méchain in 1786, but it wasn't until 1819 that the German physicist and mathematician J. F. Encke analyzed its motion, finding that it apparently broke Newton's laws of motion, including the law of universal gravitation. He discovered, in fact, that the period of motion was two hours and a half shorter than the time predicted (3.3 years) for the comet to make one complete revolution around the sun following a Kepler trajectory.

In 1868, the rotation period of Encke's Comet was still found to be slower than what Kepler's third law would have otherwise predicted, but at least the discrepancy had been reduced. Much later—in 1980—the divergence has been brought down to only a few minutes.

In the 1950s, other comets were observed and several anomalies were detected in their periods of revolution. Halley's Comet, for instance, was found to be four days late in completing its 76-year revolution around the sun.

Interested readers may consult Whipple's article, which affords details relating to the rate of decrease of the comet's mass in unit time as well as the determination of the rotation axis of the comet's nucleus and its motion of precession in space. The article also discusses the consequences concerning the thrust ($\mathbf{F}_{\text{interactions}}$ in equation [3]) exerted by sublimed gases on the comet.

In the scope of this book, I shall limit the discussion to a simple description of the phenomenon and show that it can be fully explained by applying the basic law of dynamics to Encke's Comet taken as a variable mass system.

At first glance, the comet's orbit is an ellipse with the center of the sun as one of its foci (see figure 3). From Kepler's third law the period is calculated to be 3.3 years. However, observation in 1819 revealed that the real period fell short of the predicted period by two and a half hours, which corresponds to a perturbation on the order of 1/100,000. How can such a discrepancy be explained?

Sublimation of ice on the surface of the nucleus takes place under the influence of the sun's radiation. Water vapor is ejected at an angle to the line formed from the sun to the comet, since it takes a certain amount of time for the ice to heat up (thermal inertia) and then to sublime. During that interval of time, the

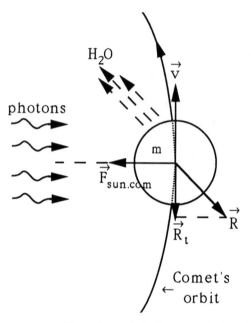

Comet's
← orbit

Figure 3. Encke's Comet

nucleus of the comet has rotated. As the water molecules are
ejected, they exert a thrust **R** on the comet which has a mass m
(figure 3). This thrust **R** constitutes an extremely weak force as
compared to the sun's force of gravitational attraction on the
comet:

$$\mathbf{R} << \mathbf{F} = - \frac{G\,M_s\,m}{r^2}\ \mathbf{e}_r\ ,$$

where M_s and m are the respective masses of the sun and the
comet, G is the constant of gravitation, and r is the distance from
the sun to the comet. At first, we can disregard **R** before **F** and
assume the comet to be in uniform circular motion. In this case,
Newton's second law will be written:

$$G\ \frac{M_s\,m}{r^2} = m\ \frac{v^2}{r}\ \text{hence: } v = \sqrt{\frac{GM_s}{r}}$$

The period is the time T it takes the comet to make one complete revolution around the sun: $T = 2\pi r/v$.

Now we shall take into account the thrust $\mathbf{R}$, which we determined by applying the law of equal action and reaction to the system composed of the comet and ejected gas during an interval of time dt. The normal component R_n of this thrust causes a very slight decrease in the sun's gravitational attraction, but this decrease has no effects on the period of the comet's orbital revolution because from one revolution to the next the situation remains stationary and the period is derived from orbital elements observed. The force resulting from solar radiation pressure on the comet—which acts in the same direction as R_n—is also disregarded. It is the influence of the tangential component R_t that becomes preponderant.

In fact, the comet's angular momentum relative to an axis Oz perpendicular to the plane shown in figure 3 and extending from the center of the sun is: $L_{oz} = mrv$ or, replacing v with its value as given above: $L_{oz} = m \sqrt{GM_s r}$. Now let us apply the angular momentum theorem at 0 (m = constant). This theorem, it will be remembered, is only a variant of Newton's second law:

$$\frac{dL_{oz}}{dt} = -r\, R_t = \tfrac{1}{2}m\, \sqrt{\frac{GM_s}{r}}\frac{dr}{dt} \, ,$$

where R_t is the tangential component of $\mathbf{R}$. This tangential component R_t tends to cause the radius of the trajectory circle (dr/dt < 0) to decrease slightly. As a result, as $v = \sqrt{GM_s/r}$, velocity will also increase slightly as well. Here we are confronted with a result which will at first seem to be a paradox: a braking force actually contributes to increase velocity of a moving object. This effect is well known in cases of artificial satellites in ''low orbit'' where they run into molecules in the upper atmosphere.

As $T = 2\pi r/v$, decreasing r and increasing v will produce a decrease in the period T. If the direction of angular rotation were reversed (see figure 3), the tangential component R_t would

31

change in sign and the opposite effect would be observed: an increase in the period as predicted by Kepler's laws. Current astronomical observations bear out such an occurrence.

The Conveyor Belt and Its Hopper

A mineral ore or grains of wheat are poured out from a hopper onto a conveyor belt moving horizontally. The ore or the wheat is carried at constant speed to a point where it is unloaded (see figure 4). Question: is it true that the conveyor belt would use no energy if there were no friction acting upon it? The answer is no. When the ore first falls from the bin, it has zero horizontal speed. However, the conveyor belt sets it in motion at the same speed at which it is moving, thus effecting a change in momentum; therefore an external force is needed and energy has to be spent.

This same type of reasoning can be found in many texts, but it lacks accuracy and is therefore a source of confusion. Some will even use this example in an effort to demonstrate that Newton's second law, if applied to a variable mass system, should be written: $d\mathbf{P}/dt = d(mv)/dt = \Sigma \mathbf{F}_{ext.}$ They maintain that the external force required to keep the conveyor belt moving at constant speed would correspond to the term $v(dm/dt)$ of the derivative with respect to time of the momentum—or change in momentum in the above answer—in a system whose mass would increase by (dm/dt) per unit of time (the fall of the mineral onto the moving conveyor belt).

However, this type of reasoning does not seem wholly satisfactory. In fact, we are now going to show, by choosing for an example a system with constant total mass, that an external force must be applied to the conveyor belt in order to keep its speed constant—thus using energy—without there being any change in linear momentum in the system chosen. To do this, let us consider

32

the upper horizontal part of the conveyor belt and the ore contained between the two dotted lines (see figure 4) as the "system."

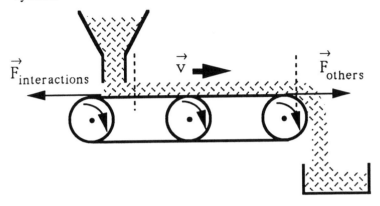

Figure 4. The Conveyor Belt and Its Hopper

At a stationary flow rate, when the belt moves at constant speed, an external force ($\mathbf{F}_{others}$ in equation [3]) must be applied to it so as to compensate for the force exerted ($\mathbf{F}_{interactions}$ in the same equation) by the ore as it falls onto the belt. To be more accurate, let us now consider all the forces acting upon our "system": the weights of both ore and conveyor belt, which are compensated for by the reactions of the wheels and supports (lack of friction and/or movement with no sliding), the horizontal force $\mathbf{F}_{interactions}$ due to the action of the ore falling from the bin (with zero horizontal speed) onto the belt and the force $\mathbf{F}_{others}$. Suppose that the collisions occurring between ore and the belt at the instant of impact are perfectly inelastic. The horizontal speed of the falling ore (dm/dt) varies suddenly from 0 to v. This transfer of momentum per unit of time corresponds to a force acting on the conveyor belt (opposite **v**) that is equal to $\mathbf{F}_{interactions} = -(dm/dt)\mathbf{v}$. This is the law of equal action and reaction forces applied to two objects: the upper left-hand part of the belt and the ore

33

falling on and making contact with the belt during an interval of time dt. Consequently, an external force $\mathbf{F}_{others}$ must be applied to our "system" with constant total mass to avoid perturbation on the state of the system, in agreement with Newton's first law on uniform rectilinear motion:

$$M \, (dv/dt) = \mathbf{F}_{interactions} + \mathbf{F}_{others} = 0$$

whence $\mathbf{F}_{others} = v(dm/dt)$.

The power imparted by the external force $\mathbf{F}_{others}$ ($\mathbf{F}_{others} \, v = v^2 dm/dt$) corresponds exactly to the work done per unit of time to increase the horizontal momentum of the ore falling on the belt at the rate of dm/dt. Such a result is in perfect agreement with the theorem of conservation of mechanical energy (the kinetic energy of our "system" is constant, $M = $ constant, $v = $ constant). Naturally, since we assumed that the collisions between belt and ore were perfectly inelastic, there will have to be heat loss corresponding to the potential energy loss of the ore in free fall from the hopper.

The objection might be raised that what is involved is a "system" with constant total mass. What happens, in fact, if we choose a variable mass "system"? For instance, the one that will come into being as soon as we start the belt going and begin to dump ore from the bin (transient state).

We shall designate $t_0 = 0$ the instant the ore first crosses the dotted line on the left of figure 4. For a certain interval of time, the system we defined above will be a "system" with variable mass $M(t)$, with the variation stemming from the increase over time in the amount of ore found between the two dotted lines in figure 4. If we apply Newton's second law in the form given by equation (3) to the system (the upper horizontal part of the conveyor belt together with the quantity of ore stated above), we obtain the same equation as we did previously:

$$M \, (dv/dt) = \mathbf{F}_{interactions} + \mathbf{F}_{others} = 0$$

since $v = $ constant, hence $dv/dt = 0$ and $\mathbf{F}_{others} = v(dm/dt)$. Assessing the external forces acting on the "variable mass system,'' we find that they are in fact the same as before. We determine these external forces—$\mathbf{F}_{interactions}$, weights, reaction of the supports—by also applying the law of equal action and reaction.

Using the basic law of dynamics in the form $d(m\mathbf{v})/dt = \Sigma \, \mathbf{F}_e$, if we consider only $\mathbf{F}_{others}$ as being an external force, leads to the proper result: $\mathbf{F}_{others} = \mathbf{v}(dm/dt)$ thanks to an error of omission: no assumption as to the collisions between ore and belt is really needed. What is more, we run into difficulties involving the theorem of conservation of mechanical energy. Indeed, those who use the formula $d(m\mathbf{v})/dt = \Sigma \, \mathbf{F}_e$ note that the power of the only external force involved is equal to two times the derivative of the variable mass system's kinetic energy $dE_c/dt = \frac{1}{2} \, v^2(dm/dt)$. They account for this difference by citing the heat losses that occur when the ore strikes the belt. Such an explanation is incorrect, since these heat losses occur even when the belt is motionless ($v = 0$) and are due to the loss of gravitational energy in the ore falling from the hopper.

A Few Other Examples and Conclusion

Another example of a variable mass system is a raindrop. As it falls, two things may happen: water vapor may condense on its surface, or evaporation can take place. Both events can lead to a change in the water drop's mass. Similarly, an elevator moving down at constant acceleration and whose mass could be changed by adding or removing elementary masses with no transfer of momentum—i.e., with no braking or no acceleration—would constitute a variable mass system to which we could apply Newton's second law ($\mathbf{F}_{interactions} = 0$ in equation [3]) in the form:

$$m(t) \, (dv/dt) = \mathbf{F}_{others} = m(t) \, g$$

Consequently, the elevator motion remains vertical and uniformly accelerated (g). This would be the case for a raindrop if isotropic evaporation occurs in space and if air resistance can be disregarded. In general, however, certain assumptions have to be made about the manner in which mass will change over time in order to determine the term $\mathbf{F}_{\text{interactions}}$ and then solve the equation of motion (equation [3]).

We hope that these simple examples have convinced the reader that Newton's second law applied to a variable mass system should be written, at any given instant t and with respect to a Galilean reference frame:

$$\Sigma \, m_i \, (dv/dt) = \Sigma \, \mathbf{F}_{\text{ext.}} = \mathbf{F}_{\text{interactions}} + \mathbf{F}_{\text{others}}$$

and not:

$$d \, \Sigma \, (m_i \mathbf{v}_i)/dt = \Sigma \, \mathbf{F}_{\text{ext.}}$$

We further hope that we have been able to persuade the reader that Newton's third law is just as basic as his other two laws. Indeed, the law of equal action and reaction forces—valid in any reference frame whether Galilean or non-Galilean—serves to determine all external forces acting on a system: $\mathbf{F}_{\text{interactions}}$ and $\mathbf{F}_{\text{others}}$.

Other advantages will result from the formulation $\Sigma m_i \, (d\mathbf{v}_i/dt) = \Sigma \, \mathbf{F}_{\text{ext.}}$ of the basic law of dynamics in systems:

1) Its universality: in this form, the law may be applied to any material system whatsoever (constant or variable total mass).
2) The law never contradicts the laws of conservation of linear and angular momenta and of energy in an isolated system.

36

Chapter 3
Inertial Forces

A reference frame in which the law of inertia—or Newton's first law—is valid is referred to as an inertia or Galilean reference frame. With respect to such a reference frame a body that is subjected to no force will remain in a state of rest or, if moving, will continue to move in a straight line at uniform velocity indefinitely.

All reference bodies in uniform rectilinear translation defined relative to a known reference body are Galilean reference bodies. Newton understood perfectly that the laws of mechanics remain identical when a transition is made from one Galilean reference body to another. According to his second law, forces are proportional to changes in velocity or, more accurately, to accelerations ($f = ma$). Accelerations have the same value in all Galilean reference frames. Consequently, Newton was able to formulate a law of Galilean relativity in the following form: "The laws of mechanics are the same in all Galilean reference frames." In other words, with mechanical experiments made within a Galilean reference frame, it is impossible to identify whether this reference frame is moving relative to another. A flight attendant inside an airplane moving at constant velocity with respect to the ground will pour coffee just as she would in her kitchen.

This conclusion involves only Galilean reference frames, for if a reference body happens to be rotating or undergoing acceleration in a straight line relative to a Galilean reference body—or if a train happens to be going around a bend, for instance—it would be impossible not to perceive the motion. What now comes into play are inertial forces due to interactions between objects. We shall now consider both the motion and the acceleration of these objects (the non-Galilean reference bodies included).

Inertial Forces: Kinematic Explanation

As we have just explained, the publication of Newton's *Philosophiae Naturalis Principia Mathematica* in 1687 made it possible to study two types of reference frames. First, Galilean reference frames with respect to which the basic law of dynamics as applied to a point mass is written:
$$\mathbf{f} = m\,\mathbf{a} \qquad (1)$$
where $\mathbf{f}$ is the sum of the forces acting on the same point mass m and $\mathbf{a}$ is its acceleration. Although acceleration is customarily assumed to be absolute, it is in fact not absolute in nature.

The second reference frame involves non-Galilean reference bodies undergoing acceleration, i.e., in any motion whatever relative to a Galilean reference body. With respect to these non-Galilean reference bodies the basic law is now written:
$$\mathbf{f} - m\,\mathbf{a}_d - m\,\mathbf{a}_c = m\,\mathbf{a}_r \qquad (2)$$
where $\mathbf{a}_r$ is the relative acceleration of mass m (with respect to the non-Galilean body considered) and $\mathbf{a}_d$ and $\mathbf{a}_c$ are the driving acceleration and the Coriolis acceleration, respectively.

Note that equations (1) and (2) are really identical, since the various accelerations are interconnected by the theorem of addition of accelerations
$$\mathbf{a} = \mathbf{a}_r + \mathbf{a}_d + \mathbf{a}_c \qquad (3)$$
Absolute acceleration = relative acceleration + driving acceleration + Coriolis acceleration. The first member of the equation is the acceleration as measured with respect to a Galilean refer-

ence body; the second member is the sum of relative, driving, and Coriolis accelerations. The driving acceleration is the acceleration of the coincident point (= same point as the moving one but assumed to be motionless with respect to the non-Galilean reference frame). The relative acceleration is measured in the non-Galilean reference frame.

If we write $\mathbf{f_0} = -m\,\mathbf{a_d} - m\,\mathbf{a_c}$, the sum of the driving and Coriolis inertial forces, then the observer on an accelerated reference body may write the basic law of dynamics in the form:

$$\mathbf{f} + \mathbf{f_0} = m\,\mathbf{a_r}, \qquad (2')$$

i.e., explicitly bringing in inertial forces.

However, the observer on an original Galilean reference body will have to write the basic law in the form of equation (1), as if the inertial forces did not exist.

Both observers have written the same basic law inasmuch as in equation (1) the inertial forces (if present) are introduced implicitly through the theorem of addition of acceleration.

Before examining the concepts of centrifugal force in particular and inertial forces in general, in connection with Einstein's equivalence principle, it is fitting to ask the following basic question: would it not be simpler to limit ourselves to only Galilean reference bodies so as to avoid bringing in the inertial forces that are the source of confusion? The answer is, unfortunately, no, for the simple reason that we are surrounded by so many accelerated rigid (material) reference bodies such as elevators, trains going around bends, rockets, artificial satellites, and even the Earth in its ceaseless rotation. In these examples it is much easier to study the motions of the objects with respect to these material reference bodies than it would be with respect to a Galilean, but immaterial, reference frame.

Indeed, although the acceleration components are concerned with purely kinematic relationships and hence do not depend on the material nature of reference bodies moving relative to one another, the laws of mechanics, on the other hand, do require material reference bodies—an elevator, a spacecraft, the Earth's surface—if only to house the observer describing the motions of objects around him. Thus we may say from now on that the terms $(-\; m\; \mathbf{a}_d\; -\; m\; \mathbf{a}_c)$ in equation (2) or $\mathbf{f}_0$ in equation (2') are physically real.

Centrifugal Force: Imaginary or Real?

The idea of a force escaping (from the Latin *fugare*: "flee") from the center has been around for a long time. Before Newton, both Descartes and Huyghens studied circular motion as dependent on the centrifugal force, which is one component of the driving inertial force introduced in the previous section. Descartes, for instance, studied the motion of a marble on the inner surface of a hollow cylinder and also that of water whirling about inside a receptacle moving in a circle. Both marble and water seemed to rush away from the center of the system. Descartes concluded that this motion resulted from the action of a centrifugal force. Today we would reason somewhat differently. In a reference frame to which the receptacle belongs, once equilibrium has been reached, the water molecules are subjected to two forces: the centrifugal force and the reaction from the walls of the receptacle. These two forces cancel each other out. On the other hand, in the reference frame of the laboratory the water molecules are subjected to only centripetal force: the reaction of the receptacle walls. This is normal, inasmuch as the water molecules move with uniform circular motion.

Many scientists have gone so far as to affirm that "it is clear today that there is no such thing as centrifugal force since this force would contradict the laws of interaction between physical objects. The illusion of centrifugal force appears whenever we view a body moving in a rotating frame."

Such views, still highly contested, are found so frequently

in textbooks that a summary of their essential conclusions is in order here. To quote one: "Inertial forces are not the result of interactions between objects, but rather of kinematic properties of non-Galilean reference bodies. One characteristic of inertial forces with respect to other forces of interaction—for instance gravific or electric—resides in the fact that there are no reactions opposing inertial forces."

In other words, the action of an inertial force being kinematic in origin, there could not be a reaction.

Some writers, on the other hand, do not hesitate to say or write that "inertial forces exist in a terrestrial reference frame, i.e., with the same rotation as the Earth's one, even if this frame were devoid of all matter."

Considering the examples involving the motion of the marble or of the water inside a receptacle as described above, we may state that such conclusions as those just quoted are contradictory, sometimes daring, outlandish, or plainly false. In the case of motion of water inside the receptacle spinning with constant angular velocity on a vertical axis, the receptacle is the non-Galilean, rigid reference body that produces the centrifugal force acting on the water molecules.

Moreover, scientists are not unanimous in accepting the arguments I have quoted above. Some, such as the one I shall quote here, have even drawn conclusions to the contrary: "It must not be said that inertial forces are fictitious. They are, it is true, purely passive. This means that inertial forces cannot cause motion, but can cause mechanical effects. They may thus cause a car to overturn, a flywheel to shatter. Or they may drive a gyroscope . . . such forces are produced or created in rigid reference bodies undergoing acceleration relative to a Galilean reference body" (2).

It is actually possible to prove the realness of these forces from Newton's first law, which states that in the absence of a force ($\mathbf{F} = 0$) an object will remain at rest (with respect to the translational motion of its center of mass). This applies to a solid sphere rotating around a diameter if it is subjected to no outside force and if the initial velocity of the center of mass is zero. One can also apply it to a point mass placed on the equator of the same sphere, even if the sphere happens to be rotating on a polar axis so that the centrifugal and gravitational forces both have the same strength. In fact, the point mass as viewed by an observer on the sphere will seem to remain in a state of rest, in equilibrium just above the surface. The point mass is thus subjected to no force ($\Sigma \mathbf{F} = 0$). We know, however, that it is indeed subjected to the sphere's gravitational attraction. This means that some real force must exist that can compensate for—or cancel—this force of gravity. This real force is the centrifugal inertial force.

Inertial Forces: Forces of Interaction between Moving Bodies

In theory it is possible to impart all types of translational and rotational motions to any rigid object. Although these motions are ordinary, they do have a remarkable feature, which we may briefy state as follows: any motion experienced by a rigid body may be broken down into two components. These are (1) a translational motion by one of its points and (2) a rotational motion on an axis passing through this point. What makes this feature so noteworthy is the fact that for a given motion, the vector of angular rotation remains the same regardless of the point that has been chosen on the rigid body to determine the translational motion. Take the example of the planets in our solar system. We know that each planet evinces two motions: translation of its center around the sun and angular rotation on its polar axis. Defining the Galilean and Copernican reference frames as we did earlier in this chapter means ascribing to them a privileged nature that allows us to simplify the expression of certain physical laws, although in so doing we obscure reality to some extent. In particular, the original statement of Newton's first law (the law of inertia) was interpreted—perhaps somewhat abusively—as applying only to a material point. However, solid objects such as the Earth or gaseous bodies such as the sun are characterized by translational and rotational motions.

One example will serve as an illustration. We know that it is simpler to describe the motion of the planet within our solar system with respect to a Galilean (or so-called Copernican) reference frame whose origin is the center of mass of the solar system and whose axes point toward fixed stars. And yet a revolving reference body connected with the sun in its own rotation might seem more practical or more valid for such a description. In this reference frame both the Coriolis and centrifugal forces acting upon the planets must be reckoned with. In any event, two observers—one within the Copernican reference frame and the other connected with the revolving sun—will find that the results involving planetary motion around the sun are in agreement. In fact, the first observer uses equation (1): $\mathbf{f} = m\,\mathbf{a}$, where $\mathbf{f}$ is the force exerted by the sun on the planet, m is the planet's mass, and $\mathbf{a}$ is the planet's acceleration measured in a Copernican reference frame. The second observer, on the other hand, uses equation (2) given above. Both, however, will have been applying the same basic law of mechanics inasmuch as the accelerations are defined by equation (3).

In this example, both centrifugal and Coriolis forces are obviously due to interactions between two different physical objects; on the one hand, the sun with its own rotation (making one complete rotation every twenty-six days near the equator, whence: $\Omega_s = 2.8\ 10^{-6}$ rad/s); on the other hand, each of the planets which we may consider as mass points, since they are so small as compared with the sun.

This interpretation naturally leads to a definition of inertial forces as being derived from the interactions of two moving bodies, with one of these bodies being a point whose own rotational motion or spin may be disregarded.

Definition of Inertial Forces

The inertial forces acting on a body assumed to be a point mass are forces of interaction between two physical objects: the point mass and a solid (or gaseous) body moving in any manner with respect to a Galilean reference frame. These interaction forces arise from the accelerated motions of the solid or gaseous body (i.e., translation and rotation) and/or from the relative motion of both objects one with respect to the other. However, the exact way in which they arise remains unexplained.

The above definition offers a certain number of advantages in that:

1) it is in perfect agreement with Einstein's equivalence principle;
2) it allows us to show that inertial forces obey Newton's (generalized) third law;
3) it shows that Newton's three laws of motion are closely connected to the three principles of conservation (linear and angular momentum and energy) in an isolated system;
4) it does not stand in contradiction to long-observed phenomena such as the flattening of the planets in our solar system or the tides and the various Coriolis effects on the Earth's surface.

In fact, most of the planets in the solar system, including the Earth, have the shape of a revolving ellipsoid whose smaller axis (or polar axis) is the plane's very rotational axis. The flattening near the poles and the equatorial bulge are due to the action of centrifugal forces on the molecules of matter of which the planet is composed.

Tides, on the other hand, arise from inertial forces that are themselves spawned by the motions of the Earth around the sun and the moon around the Earth. This phenomenon leads to tidal bulges on the side of the Earth under and the side of the Earth opposite the attracting body (sun or moon). The tidal effects produced by the sun are weaker than those caused by the moon (in a proportion of $\frac{1}{3}$ to $\frac{2}{3}$). The shift in the areas of maximum bulge are due mainly to the Earth's diurnal motion, i.e., the earth's rotation once every twenty-four hours. This means that tidal occurrences are most pronounced whenever the celestial body exerting the attractive force happens to be in the plane of the equator. What is more, the effects of the sun and the moon are compounded when these two bodies are in opposition or conjunction. They are weaker when the sun and the moon are in quadrature. Tides are said to be weakest when the moon is in first or last quarter (neap tide), greatest at full or new moon (syzygy, spring tides). Tides are particularly noticeable when full and new moons occur near the equinoxes.

Let us now look at some instances of the Coriolis effect as it might occur at the Earth's surface. A cannonball shot toward the north from a point in the northern hemisphere will be deflected toward the east. At our latitudes in the northern hemisphere the prevailing winds are from the west, following isobars as they blow (rather than blowing toward low pressure areas). The forces due to pressure gradients are compensated by Coriolis forces. Finally, Coriolis forces are responsible for the twisting of winds that gives rise to cyclones and hurricanes.

To sum up, we have attempted to provide an answer to the question as to the origin of inertial forces. In the examples we have cited, we noted that inertial forces were forces of interaction between two physical bodies in motion: on the one hand the sun, the Earth, or another planet and on the other hand an object considered as a point, such as a shell, water molecules in the oceans, air molecules in the atmosphere, etc.

One question naturally remains: how can a kinematic transformation create inertial forces? Or still, why do inertial forces adapt themselves so well to such a kinematic transformation (the theorem of acceleration components)? Are we confronted with a mystery here, or might there be a connection with Einstein's equivalence principle, which says that gravitational and inertial masses are equal?

Einstein's Equivalence Principle

Since Newton's time it has been known that gravitational mass and inertial mass have one and the same value. The gravitational mass m_g of a particle (or body) is, it will be remembered, the mass found in the gravitational force exerted by another mass and the inertial mass is found in the basic law of dynamics ($f = m_i a$). Newton himself was intrigued by this equivalence, but did not venture to explain it.

Nevertheless, it is this equivalence that allows a material point to be in equilibrium relative to a sphere rotating around its diameter. (Cf. the end of the section titled "Centrifugal Force: Imaginary or Real.") This eqivalence of the two masses also makes it possible for an astronaut to experience weightlessness inside his space vehicle. If the ratio m_g/m_i depended solely on the nature of the objects, wooden objects, for instance, might be weightless while those made of steel would not be. In the event of sudden acceleration, such a state could lead to severe internal tension in either the matter itself or the various objects found within non-Galilean material reference frames. These observations apply to Coriolis effects and, in general, to all effects caused by the accelerated motions of material reference bodies, whether a gravitational field is present or not.

Einstein was the first to consider this equivalence as a basic principle or postulate. A simplified statement might read as follows: "We have equivalent points of view on gravity within a Galilean reference frame where there is a uniform gravitational field or in an accelerated reference frame relative to the previous one but where gravity does not exist as a force. Gravitational mass and inertial mass now appear as one and the same."

In the preceding statement, the non-Galilean reference frame being considered must be undergoing constant acceleration a_d relative to a Copernican reference frame, as in the case of a space vehicle that is being propelled by a rocket far from any mass in the universe. Inside the vehicle an object with mass m is submitted to the driving inertial force—ma_d (similar to a weight m**g**, with **g** being a constant). This force arises from the uniformly accelerated motion of the space vehicle.

In other words, according to Einstein, the inertial forces are similar to the forces of gravity. As we saw previously, this viewpoint does not contradict the law of inertia as generalized so as to apply to motions of uniform rotation of solid objects. To make the equivalence and/or similarity between gravity and inertial forces clearer, we shall show in several classic examples that the forces of inertia also obey the generalized law of equal action and reaction (Newton's third law).

Objects Falling in a Moving Elevator

Let us now consider the thought experiment devised by Einstein and Infeld (3) involving two objects: a handkerchief and a watch. These two objects will be imagined as falling in an elevator, which is itself in a state of free fall in space as it falls toward the Earth's surface.

To the observer outside, the handkerchief and watch both fall with the same acceleration. The same also holds true for the elevator, its walls, its floor, and its ceiling. Hence the distance between the two objects and the floor will not be seen to vary.

But to an observer inside the elevator, the two objects will remain exactly where they were when he dropped them. He can disregard the gravitational field, since its cause is found outside his reference system. He thus finds that inside the elevator there is no force acting upon the two objects that appear to be at rest, just as if they were in a Galilean reference frame. Should the observer impart to one of the objects a motion in any direction, up or down, for instance, it will still move uniformly as long as it does not come into contact with the elevator ceiling or floor. Briefly stated, this means that the laws of classical mechanics are valid for the observer inside the elevator. All bodies will be found to behave in accordance with the law of inertia.

Following this description and a few other considerations, Einstein and Infeld introduce the principle of equivalence of gravitational mass and inertial mass. This principle reconciles the viewpoints of both inside and outside observers.

Let us at this point attempt to account for this thought experiment in terms of interactions between the various objects involved: the Earth, the elevator, and the handkerchief or watch, labeled M_{ea}, M_{el}, and m, respectively. The law of equal action and reaction applies only to the interaction occurring between two bodies. In this example, three independent interactions must be reckoned with:

1) the Earth-elevator interaction $(M_{ea} - M_{el})$
2) the Earth-handkerchief interaction $(M_{ea} - m)$
3) the elevator-handkerchief interaction $(M_{el} - m)$.

The first two interactions are classic ones involving a gravitational interaction as described by the law of universal gravity. In this case we know that if the Earth attracts the elevator (or the handkerchief, $M_{el}g$ or mg), the elevator (or handkerchief) also attracts the Earth with an equal but opposite force. The third interaction, however, involving the elevator and the handkerchief, cannot be gravitational in nature inasmuch as the handker-chief (or the watch) is found inside the elevator. According to Gauss's theorem, the force of gravitational interaction (M_{el} − m) is zero. There remains the inertial force due to the accelerated motion of the elevator with respect to the surface of the Earth ($\mathbf{a}_d$ = $\mathbf{g}$). This force, $-m\mathbf{a}_d = -m\mathbf{g}$, is equal and opposite to the weight of the handkerchief m$\mathbf{g}$. Hence, for the observer inside the elevator, the handkerchief remains in its state of rest, since the sum of the forces exerted on it by the elevator undergoing acceleration and by the Earth is zero. To the action represented by the inertial force ($-m\mathbf{a}_d$) there corresponds a reaction ($+m\mathbf{a}_d$ = m$\mathbf{g}$) exerted on the elevator, with both forces being evaluated by the observer inside the elevator if he knows that his elevator is undergoing uniform acceleration toward the Earth's surface. Should he ignore that his elevator is moving in a uniform gravita-tional field, he will say that the elevator–handkerchief action and reaction are null. But the observer outside the elevator will not see any contradiction, since what he sees is an elevator, and everything inside it, moving with uniform acceleration toward the Earth's surface: (M_{el} + m)$\mathbf{g}$.

To recapitulate: taking into account the principle of equiva-lence of gravitational mass and inertial mass, we note in this example that the gravitational forces and the inertial forces both obey in one way or another Newton's three laws of motion.

The Coriolis Force: Easterly Deflection of an Object Falling Freely toward the Surface of the Earth

Consider a marble with mass m, which, because of its small size when compared with the Earth's, may be assumed to be a point mass. The Earth rotates on its polar axis ($\omega/2\pi$ = 1 com-plete rotation every twenty-four hours). To simplify matters, we shall place ourselves at the equator and call A the point found at altitude h above the ground where the marble with mass m will be dropped without initial velocity (see figure 5). Now let O be the point on the ground on the vertical line passing through A. The marble's free fall can be studied either in a Galilean reference frame (i.e., outside the Earth's rotational motion) or in a non-Galilean reference frame attached to the Earth.

In the latter instance, and disregarding Coriolis effects, we are confronted with the classic problem of a body moving freely in a uniform gravitational field inasmuch as the marble is subjected only to its weight: **mg**. Its motion is therefore uniformly accelerated even if the acceleration field of weight, at the equator, is weakened by the centrifugal acceleration field caused by the earth's own rotation. As the marble moves freely, the law of equal action and reaction is respected at every instant. If the Earth attracts the marble (action **mg**, including gravitational and centrifugal forces), the marble will also attract the Earth with an equal but opposite force: the reaction −**mg**.

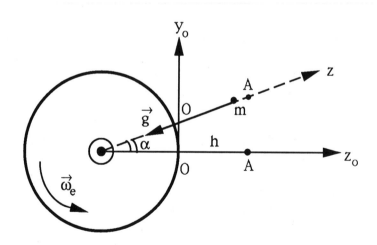

Figure 5. A Marble Falling Freely toward the Earth

Actually, even if Coriolis effects can often be disregarded, they do exist. In the case of the marble in free fall, this effect appears as an easterly deflection (direction $0y_0$ in figure 5) of the point of impact. There is, in fact, a noncentral inertial force which acts upon the marble in this direction. This force, proportional to the marble's velocity as it falls, is the Coriolis force F_c due to the interaction of two moving objects: the marble, which is moving with constant acceleration toward the ground, and the Earth, which is rotating on its polar axis. If the Coriolis force F_c exerted by the Earth on the marble obeyed Newton's (generalized) third law, the observer on the Earth would conclude that the marble exerts an equal and opposite force on the Earth ($-F_c$ applied to the center) and at the same time a couple, making it possible to compensate for the moment of the Coriolis force relative to the center of the Earth.

Of course, the observer on the Earth has no proof that what he says is correct. On the other hand, an observer connected to a Galilean reference frame ($0x_0y_0z_0$ as determined by the position of $0Az_0$ of the vertical line when the marble is dropped and outside the Earth's rotational motion) will be able to help him out. In fact, this observer does not have to reckon with inertial forces, since he finds himself in a Galilean reference frame. The force acting upon the marble is the force of gravity mg (see figure 5) which may be broken down into two terms: $-$mg cosα $\mathbf{e}_{z0}$ and $-$mg sinα $\mathbf{e}_{y0}$ = $-$mg $y/R_e\mathbf{e}_{y0}$. As at first glance cos α = 1, the marble's motion as seen in $0x_0y_0z_0$ has two components: free fall along $\mathbf{e}_{z0}$ and a slightly decelerated transversal motion along $\mathbf{e}_{y0}$.

When the marble reaches the ground, the result as witnessed by both observers is the same. The point of impact has been deflected toward the east with respect to the foot of a vertical plumbline.

For the observer in the Galilean reference frame $0x_0y_0z_0$, the force $-$mg sinα $\mathbf{e}_{y0}$ (which is a gravitational force) is the physical agent causing the marble to decelerate in the direction $\mathbf{e}_{yo}$ (in an easterly direction). In fact, the marble must undergo deceleration in its transversal motion; otherwise it would travel too far with respect to the foot $0'$ ($0'$ being motionless with respect to the Earth, 0 being the fixed origin of the Galilean reference frame once the marble has been dropped). This would happen since the marble's initial velocity was greater than the velocity of point $0'$ with respect to $0x_oy_oz_o$. The Galilean observer would thus note that a force gravitational in origin is equivalent (or similar) to the Coriolis inertial force used by his terrestrial colleague. These forces causes the Coriolis effect, characterized by an easterly deflection of the marble's point of impact. This phenomenon can be observed in both Galilean and non-Galilean reference frames. The two forces, $\mathbf{F}_c$ and $-\mathbf{mg} \sin\alpha \, \mathbf{e}_{yo}$, are opposite in direction, a fact explained by the relativity of the motions. If the Earth moves with an angular rotation ω_e with respect to a Copernican reference frame, then the Copernican reference frame moves with angular rotation $-\omega_e$ with respect to the Earth.

Finally, the observer in the Galilean reference frame $0x_oy_oz_o$ can verify that the force of gravitational attraction exerted by the Earth on the marble ($-\mathbf{mg} \cos\alpha \, \mathbf{e}_{zo}$ and $-\mathbf{mg} \sin\alpha \, \mathbf{e}_{yo}$), obeys Newton's (generalized) third law. The fact that the $-\mathbf{mg} \sin\alpha \, \mathbf{e}_{yo}$ component exists shows that a gravitational force can be noncentral in nature. In accordance with the generalized law of action and reaction (see chapters 1 and 4), to the force $-\mathbf{mg} \sin\alpha \, \mathbf{e}_{yo}$ there must be an equal and opposite corresponding force, $+\mathbf{mg} \sin\alpha \, \mathbf{e}_{yo}$, exerted by the marble on the center of the Earth, as well as a couple compensating for the moment of the force $-\mathbf{mg} \sin\alpha \, \mathbf{e}_{yo}$ measured at the center of the Earth. This couple, which is exerted by the marble on the Earth, tends to accelerate the Earth's rotation.

The Galilean observer can prove what he asserts in several ways. First of all, he can apply the theorem of kinetic energy to the marble between t_o, the instant the marble is dropped, and t_f, the instant it reaches the ground. In doing so he verifies that the work of the force $-mg \sin\alpha$ really does correspond to the variation in the marble's transversal kinetic energy between instants t_0 and t_f. He will also note that it is the component $-mg \sin\alpha$ of the force of gravity that causes a decrease in the marble's angular momentum with respect to the Earth's polar axis. And last, as the Earth–marble system may be considered as an isolated system, the total angular momentum of the system relative to the polar axis must remain constant. As the marble's angular momentum decreases as it falls, the angular momentum of the Earth's rotation must increase in the same amount. This can happen only by means of the couple exerted by the marble on the Earth.

Finally, the interested reader can verify that the law of conservation of energy is respected in the Galilean reference frame whose origin is the Earth's center, as well as in the non-Galilean reference frame that is the surface of the Earth. In fact, in both reference frames the energy source is the same: the initial gravitational potential energy mgh (Earth–marble).

This example involving the easterly deflection of a marble moving freely over the Earth's surface has brought to light the fact that the Coriolis inertial force—i.e., the noncentral force exerted by the Earth on the marble—was similar (or equivalent) to a force of gravity. In addition, the "Coriolis effect," meaning the selfsame easterly deflection, will be observed regardless of the reference frame, Galilean or non-Galilean, in which the observer finds himself. The interactions occurring between physical bodies such as the Earth–marble system can take place not only through equal and opposite central forces ($-m\mathbf{g}$ and $+m\mathbf{g}$) but also through couples whenever the interaction forces are noncentral. Such conclusions are borne out both by Einstein's equivalence principle (gravitational mass = inertial mass) and by the laws of conservation (linear and angular momentum and energy) in an isolated system.

Chapter 4
Newton's Third Law (Generalized)

Many scientists believe that Newton's third law, the law of action and reaction, is not universally respected. To prove this, they often cite instances involving relativistic occurrences such as the case of magnetic forces acting among moving charged particles. In this example Newton's third law is broken, they say, due to the finite value of the speed of the signals conveying the interactions. Some scientists even cite examples drawn from classical physics (v $<<$ c): the law of action and reaction would not be obeyed in the case of the electric dipole-point charge interaction or in a dipole-dipole interaction. The law would also be inapplicable to inertial and Coriolis forces. However, we demonstrated in the previous chapter that such was not the case.

Nevertheless, all scientists are unanimous in asserting that the laws of conservation of linear and angular momentum and of energy in an isolated system cannot admit a single exception, whatever the field of physics in question may be (see chapter 1). Now there would seem to be a contradiction between the two statements: no violation in the law of conservation of momentum can be admitted, while the law of equal action and reaction, particularly in electrostatics, may at times be broken. In fact, if Newton's third law is valid, then it follows that momentum must be conserved. The inverse is also true: if Newton's third law were really violated, then we should be able to detect anomalies in the conservation of linear and angular momentum and of energy in an isolated system.

K. R. Symon (4) suggests two possible interpretations of Newton's third law as originally stated (set forth in chapter 1). The first interpretation, or *strong form*, would require that action and reaction not only be equal and opposite, but that they be shown by a straight line connecting the particles involved. The second interpretation, or *weak form*, would not stipulate that interaction forces must be central as long as one equivalence were maintained: $F_{12} = -F_{21}$. To illustrate this weak interpretation, I would cite the magnetic interaction which takes place between a magnet and a charged particle moving at velocity v near the magnet's north pole. These suggestions, however interesting they might be, are nevertheless inadequate, inasmuch as we must also reckon with the appearance of couples in instances where noncentral interaction forces exist. Hence we are led to formulate a more general statement of Newton's third law (see the statement provided in chapter 1).

The Generalized Law of Equal Action and Reaction

In each instance where body A exerts an action on body B, body B will exert an equal but opposite reaction on body A. The action (or reaction) of one body on the other consists of a combination of forces that at a point 0 in space are reduced to a vector sum F_{AB} and to a resultant moment that itself is defined by the theorem of the transport of moments:

$$M_{AB} = \Gamma_{AB} + OB*F_{AB}$$

where F_{AB} is the force exerted by body A on body B, Γ_{AB} is the couple, whose axis passes through B, exerted by A on B, and the term $OB*F_{AB}$ is the moment at point 0 of the force F_{AB}.

Mathematically, the generalized law of equal action and reaction is expressed by means of the two following vector equations:

$$F_{AB} + F_{BA} = 0 \qquad (1)$$

and: $$(M_{AB})_o = -(M_{BA})_o \qquad (2)$$

Forces $\mathbf{F}_{AB}$ and $\mathbf{F}_{BA}$ may or may not have the same line of action. If they do not have the same line of action, then there must needs be an interaction by means of couples as in the case of the point charge-dipole interaction (see the following section). If forces $\mathbf{F}_{AB}$ and $\mathbf{F}_{BA}$ do have the same line of action, we find the classic statement of the law of equal action and reaction; the couples Γ_{AB} and Γ_{BA} are null, and the interaction forces are central.

This law is general and can be applied in this form even if the two bodies (or one of them) are point particles. For instance, we noted in the preceding chapter that the marble falling toward the ground exerts on the Earth a couple that tends to accelerate the rotational motion of the Earth on its axis, a phenomenon observed in a Galilean reference frame.

Our generalized statement of the law of equal action and reaction does not contradict Newton's original statement but adds to it. It can thus predict interactions occurring between physical objects through noncentral forces and couples—i.e., electromagnetic forces, Coriolis inertial forces, and even certain forces that are gravitational in origin, as we shall see in the case relating to the (retrograde) precession of the equinoxes. *From this statement we may deduce the laws of conservation of linear and angular momentum in an isolated system.* In this form, the law of equal action and reaction has probably not been violated. This is what we shall demonstrate in the examples that we described at the beginning of this chapter.

Point Charge-Electric Dipole and Dipole-Dipole Interactions

The model of an electric dipole ($\mathbf{p} = Q \mathbf{d}$) is generally represented by two charged particles ($-Q$ and $+Q$) placed at a distance d from each other. Such a dipole constitutes an object, and if we consider its electrostatic interaction with a point charge $+q$ placed at a distance from it, we note that the law of equal action and reaction as expressed in the strong form (i.e., equal and opposite central forces) is most often violated. Certain scientists will explain that in this case the point charge-dipole electrostatic interaction must be broken down into three Coulomb interactions: $(-Q, +Q)$, $(-Q, +q)$, and $(+Q, +q)$. They go on to say that for these last three interactions between point charges, the law of equal action and reaction as given in its strong form is adequate.

In chapter 6, we shall derive from electrostatic laws the expressions of the forces exerted by the dipole **p** on the charge q and the forces exerted by the charge q on the dipole **p**, as well as the expression of the couple Γ_{qp} exerted by the charge q on the dipole **p** as pertaining to a relative dipole-particular charge arrangement. This will enable us to verify that equations (1) and (2) given above are adequate:

$$\mathbf{F}_{pq} + \mathbf{F}_{qp} = 0 \qquad (1')$$

and: $$(\mathbf{M}_{pq})_B = -(\mathbf{M}_{qp})_B \qquad (2')$$

where B is the position of the point charge q,

$$(\mathbf{M}_{pq})_B = 0 \qquad \text{and} \qquad (\mathbf{M}_{qp})_B = \Gamma_{qp} + \mathbf{BO} * \mathbf{F}_{qp}$$

0 is the center of the dipole.

Such a result is a general one. Whatever the position of the point charge with respect to the dipole, the mutual actions of both bodies—dipole and charge—obey the generalized law of equal action and reaction.

The same type of calculation may be made in the case of the interaction between two electric dipoles. These calculations might appear overly complex, and so we shall not get into them at this point. We must stress, however, that our preceding conclusions remain valid: regardless of the positions and relative orientations of the dipoles, the mutual electrostatic actions of both bodies (the dipoles) will obey the generalized law of equal action and reaction. These conclusions are in perfect accordance with the concept of interaction potential energy of the system made up of two dipoles. The presence of interaction couples is obvious, since in keeping with electrostatic laws we know that a dipole inserted in an electric field $\mathbf{E}$, created, for instance, by the other dipole, is subjected to a couple $\Gamma = \mathbf{p} * \mathbf{E}$, the vector product of dipole $\mathbf{p}$ and field $\mathbf{E}$.

Magnetic Interactions: Magnetic Force

Magnetic interactions are perhaps somewhat more delicate to describe inasmuch as they bring in relativistic phenomena. Some scientists go so far to assert that a force depending on speed (as is the case with magnetic force or Coriolis force) cannot obey the law of equal action and reaction because the signals conveying the interactions travel at finite speed. However, if we limit ourselves to the field of classic electromagnetism as defined by Maxwell's equations, we shall be able to demonstrate that in classic examples the electromagnetic interaction forces between two physical objects obey Newton's (generalized) third law. Before turning to this demonstration, it is important to recall a few definitions involving "electromagnetic effects."

A point particle q moving in an electromagnetic field ($\mathbf{E}$, $\mathbf{B}$) is subjected to two forces: (1) an electric force $\mathbf{F}_{el} = q\mathbf{E}$, with $\mathbf{E} = -\vec{\nabla}V - \partial\mathbf{A}/\partial t$, where V is the electric potential and $\mathbf{A}$ is the magnetic vector potential prevailing in the vicinity of the particle, and (2) a magnetic force $\mathbf{F}_m = q\mathbf{v} * \mathbf{B}$. This magnetic force is the vector product of the velocity $\mathbf{v}$ of the charge q and the magnetic field $\mathbf{B}$. These quantities ($\mathbf{v}$, $\mathbf{E}$, $\mathbf{B}$, V and $\mathbf{A}$) are all evaluated with respect to a Galilean frame chosen for the purpose of studying the electromagnetic interaction under consideration. If only magnetic interactions are involved, then the electric potential V is zero, but the term $-\partial\mathbf{A}/\partial t$ may not be null, even if a magnetostatic field is involved. This is due to the motion of the charged particle q in the magnetic field $\mathbf{B}$.

Example 1: Definition of the Ampere

For this example we shall consider two straight parallel wires of the same length L through which the same current I passes and whose distance d is much shorter than L. Applying the formula which gives the magnetic field created by an indefinite rectilinear current and Laplace's law (relating to the action of a magnetic field on a current), we can demonstrate that the two wires attract each other with the force as shown in the following equation:

$$f = \frac{\mu_0}{2\pi} \frac{I^2}{d} \, L$$

In the MKSA (meter, kilometer, second, ampere) system of units the standard ampere is defined as the intensity that, when flowing through the type of system as we have just described, will produce a force per unit of length (f/L) of 2.10^{-7} newtons per meter when $d = 1$ meter. In this example, the force exerted by wire 1 on wire 2 is equal to but opposite that exerted by wire 2 on wire 1.

Example 2: The Interaction of an Indefinite Straight Wire through Which a Current I Flows and a Charge q Moving at a Velocity v

Under the assumption that $q > 0$, three different situations are encountered, depending on whether the velocity **v** of the charged particle q is parallel or perpendicular to the straight wire.

In the first case, the velocity **v** of the charge q is parallel to the rectilinear current I and in the same direction. It can be easily shown that the current I and the charge q attract each other mutually with the force:

$$f = \frac{1}{2\pi\epsilon_0 c^2} \frac{Iqv}{r} \, ,$$

where r is the distance separating the charge from the wire, ϵ_0 is the electromagnetism constant, and c is the speed of light. Here, once again, the law of equal action and reaction is respected in its simplest form.

In the second case, the velocity **v** of the charged particle is perpendicular to the current I, along a line of force of the magnetic field created by the current. Here we see immediately that action and reaction are both null (hence equal and opposite).

In the third case, the velocity **v** of the charged particle is perpendicular to the rectilinear current I in the plane P defined by the same current and the point charge q. The calculations involved here are far from simple, but by using the theory of special relativity it can be shown that the action of the wire, through which a current I flows, on the charge is represented by the noncentral magnetic force (since it is parallel to the wire) and that the reaction exerted by charge q on the wire is a force equal to and opposite the first force at the point of intersection between the rectilinear current with the support of **v** and a couple, the axis of which is perpendicular to plane P. In this third instance, the generalized law of equal action and reaction is respected.

Example 3: The Interaction between a Circular Coil through Which a Current i Flows and a Charge q Moving at a Velocity v

This example of magnetic interaction shows that the charged particle q must be subjected to not only a magnetic force ($\mathbf{F}_m = q\mathbf{v} * \mathbf{B}$) but also an electric force linked to the variation in the magnetic vector potential **A**. This is a variation that is experienced by the particle moving at velocity **v**. In other words, the total force exerted by current i on charged particle q must be equal to:

$$\mathbf{F}_{iq} = \mathbf{F}_m - q \frac{\partial \mathbf{A}}{\partial t}$$

Calculations will show that the generalized law of equal action and reaction is indeed respected:

$$\mathbf{F}_{iq} = -\mathbf{F}_{qi} \qquad \text{and} \qquad (\mathbf{M}_{iq})_H = -(\mathbf{M}_{qi})_H$$

where H is the point at which the particle is found at a given instant t.

We could go on with more examples, but it would be overly fastidious to do so. We shall nevertheless point out that the interactions between two dipoles may be treated in the same way we deal with interactions between two electric dipoles. The conclusions will be identical: Newton's (generalized) third law will be respected.

Precession of the Equinoxes

The center (O) of the Earth describes a fairly circular path around the sun (the period being one year) in the plane of the ecliptic. At the same time, the Earth rotates on a polar axis (the period being twenty-four hours). This axis undergoes a slow shift known as precession: the Earth's axis slides along the surface of an imaginary cone, perpendicular to the ecliptic and with a half angle at its apex of 23.5°. The precessional motion is exceedingly slow; one complete cycle of the axis as it describes the conical motion requires twenty-six thousand years. As the polar axis shifts because of precession, the equinoxes (the points of intersection of the celestial equator and the ecliptic) slide around the sky. This westward movement of the equinoxes along the ecliptic was discovered by Hipparchus about 130 B.C.

As Newton suggested as early as 1687, this gyroscopic effect may be explain by the appearance of two couples exerted on the Earth: one by the sun and the other by the moon. Indeed, periodically (twice a year for the sun and twice each lunar month for the moon) the forces exerted by the attracting celestial body on the equatorial bulge of the Earth act vectorially in such a way that at the center (O) of the Earth there appears a couple with an axis perpendicular to the Earth's polar axis and oriented along the line of equinoxes. The effect of such couples is to bring the Earth's equatorial plane back into the plane of the ecliptic, thereby causing the average motion of precession in question.

The Method for Determining the $\Gamma_{sun-earth}$ Couple (at Winter or Summer Solstices)

Thanks to Newton's (generalized) third law, we can now propose a method for calculating these couples. For instance, let us take the case of the sun–Earth interaction, as it occurs at the time of the winter solstice. In our example we shall refer to the sun as body A and to the Earth as body B.

At this particular time of the year, the Earth's polar axis is in the plane normal to the ecliptic passing through the center (O_1) of the sun. In figure 6 we show the Earth as an ellipsoid flattened at the poles and submitted to only the attraction of the sun which we here assume to be perfectly spherical. Oz is the normal to the

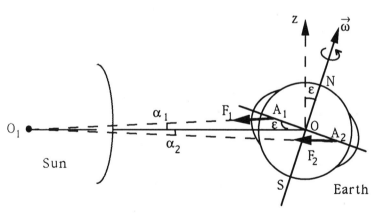

Figure 6. Vectorial Addition of the Forces Exerted by the Sun on the Earth

ecliptic; SN is the Earth's polar axis and ϵ is the inclination of the Earth's equator to the ecliptic. The plane perpendicular to plane O_1Oz, represented by the SN axis, divides the Earth's equatorial bulge (R_{ee} = 6,378 km. at the equator, R_{ep} = 6,357 km. at the poles) into two parts, with one of these parts D_1 being closer to the sun than the other part, D_2. The centers of mass A_1 and A_2 of part D_1 (left) and D_2 (right) of the bulge have also been shown in the figure.

The forces of attraction exerted by the sun on the Earth may be broken down into three forces: $\mathbf{F}$, $\mathbf{F_1}$ and $\mathbf{F_2}$. Force $\mathbf{F}$, not shown in figure 6, applied at the center of the Earth (O) and oriented toward the center of the sun (O_1) represents the overall gravitational attractions of all the mass elements found within the sphere having a radius R_{ep}. As for $\mathbf{F_1}$ and $\mathbf{F_2}$, they are respectively those forces exerted by the sun on the two parts of the equatorial bulge, i.e., D_1 (on the left) and D_2 (on the right). These forces are applied at points A_1 and A_2 as shown in the following mathematics:

$$\mathbf{F_1} = - \frac{G\ M_s\ m}{2[r^2 + OA_1^2 - 2r\ OA_1 \cos\epsilon]}\ \frac{O_1A_1}{O_1A_1}$$

$$\mathbf{F_2} = - \frac{G\ M_s\ m}{2[r^2 + OA_2^2 + 2r\ OA_2 \cos\epsilon]}\ \frac{O_1A_2}{O_1A_2}$$

where G is the constant of universal gravitational attraction, M_s the mass of the sun and m the mass of the Earth's equatorial bulge.

Vectorial reduction at 0 of the three forces, $\mathbf{F}$, $\mathbf{F_1}$ and $\mathbf{F_2}$ gives:

(a) a central force

$$\mathbf{F}_{sun-earth} = - \frac{G\ M_s\ M_e}{r^2}\ \frac{O_1O}{O_1O}$$

where M_e is the mass of the Earth and $O_1O = r$ is the distance from the sun to the Earth.

(b) a noncentral force, $\mathbf{f}_{sun-earth}$ = projection of the force vectors $\mathbf{F_1}$ and $\mathbf{F_2}$ onto the Oz axis, shown algebraically as:

$$\mathbf{f}_{sun-earth} = -(F_1 \sin \alpha_1 - F_2 \sin \alpha_2) \, \mathbf{e}_z \qquad (3)$$

(c) a couple, $\Gamma_{sun-earth}$ that tends to bring the Earth's equatorial plane back into the plane of the ecliptic.

In accordance with the generalized law of equal action and reaction, the following equations must result:

$$\mathbf{F}_{sun-earth} + \mathbf{f}_{sun-earth} = -B\mathbf{F}_{earth-sun} - \mathbf{f}_{earth-sun}$$
$$\text{and:} \qquad (\mathbf{M}_{sun-earth})_{O1} = -(\mathbf{M}_{earth-sun})_{O1}$$

where the resulting moments are evaluated, for instance, at the center of the sun, O_1. Assuming that the sun is spherical enables us to state that the couple $\Gamma_{earth-sun}$ exerted by the Earth on the sun is null, and, further, that the resulting moment $(\mathbf{M}_{earth-sun})_{O1}$ of the Earth's action on the sun must needs be null. The couple $\Gamma_{sun-earth}$ exerted by the sun on the Earth is thus defined by the vectorial equation:

$$(\mathbf{M}_{sun-earth})_{O1} = 0 = \Gamma_{sun-earth} + \mathbf{O}_1\mathbf{O} * \mathbf{f}_{sun-earth} \qquad (4)$$

Equation (4) together with a similar equation in the case of the moon–Earth interaction, make it possible to evaluate numerically couples $\Gamma_{sun-earth}$ and $\Gamma_{moon-earth}$ exerted by the sun and the moon on the Earth when these couples reach their greatest strength. In an effort to keep within the scope of this book, and in view of the fact that similar calculations of these couples, although based on other methods, exist in the scientific literature, we shall only say here that the numerical values which we obtained using equations (3) and (4) above agree with the results obtained through astronomical observations. (ω_e = the Earth's rotational speed; Ω_{prec} = the angular velocity of the precession of the equinoxes.) In addition, the values we calculated in this manner are also found to agree with the measured values of other constants: G = the universal constant gravitation, M_s, M_e, M_m represent the masses of the sun, Earth and moon respectively, and R_{ee} and R_{ep} stand for the radii of the Earth at the equator and the poles respectively.

These results have certainly been known to astronomers for a long time, inasmuch as the effects of these various interactions are included in the calculations used in determining the motions of the planets in their revolution around the sun. And yet to our knowledge no writer has ever mentioned both the orthogonal component of the sun (sphere)–Earth (flattened at the poles) force of interaction and the $\Gamma_{sun-earth}$ couple. However, in the framework of this study concerning the law of equal action and reaction, this component of a noncentral force, gravitational in origin, plays an important part, even though it is not very strong. The reader can convince himself that this force exists by referring to figure 6. It must be pointed out, however, that this noncentral force (equation [3]) changes direction at every half-period. At the winter solstice it is oriented toward $z < 0$, but at the summer solstice it has shifted in the opposite direction, toward $z > 0$. The $\Gamma_{sun-earth}$ couple, on the other hand, always tends to bring the earth's equatorial plane back toward the plane of the ecliptic in accordance with the vectorial equation (4).

In recapitulation, this section has set out to show that Newton's (generalized) third law applies to the Earth–sun and Earth–moon forces and couples of gravitational interactions. Proof of this may be found in the detailed study of the Earth's motion relating to precession of the equinoxes. The findings reveal that gravitational interactions between two physical bodies (sun–Earth) can occur by means of noncentral forces even if the forces of attraction between two point masses are central.

We also saw that similar conclusions could be drawn about electrostatic and magnetic interactions, as well as those interactions taking place by means of inertial forces (since every interaction implies the presence of two physical bodies). All these forces of interaction behave in the same manner with respect to Newton's (generalized) three laws.

Eletromagnetic Interactions between Two Moving Charged Particles

As we mentioned at the beginning of this chapter, many scientists believe that the forces which are exerted between two moving charged particles do not as a general rule obey Newton's third law. The example which is most often cited in the literature concerns two charges that are moving at a given instant, following paths that are perpendicular to each other and found within the same plane. This situation is shown in figure 7.

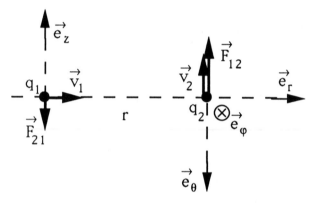

Figure 7. Two Moving Charged Particles with Mutually Perpendicular Velocities at Time t

Thus the charge q_2, moving at a velocity $\mathbf{v}_2 = -v_2\mathbf{e}_z$, gives rise to a magnetic field at the location of q_1:

$$\mathbf{B}_2 = \frac{\mathbf{v}_2 * \mathbf{E}_2}{c^2} = \frac{-v_2 q_2}{4\pi\epsilon_0 \, c^2 r^2} \, \mathbf{e}_\phi;$$

this magnetic field exerts a magnetic force on particle q_1:

$$\mathbf{F}_{m21} = q_1 \, \mathbf{v}_1 * \mathbf{B}_2 = \frac{v_1 q_1 \, v_2 q_2}{4\pi\epsilon_0 \, c^2 r^2} \, \mathbf{e}_\theta.$$

As the magnetic field created by particle q_1 (moving at a velocity

66

$\mathbf{v}_1 = v_1 \mathbf{e}_r$) in the area where q_2 is found is null, the magnetic force exerted by q_1 on q_2 is equally null. However, in this case, the action of q_1 moving at a velocity v_1 on q_2 moving at a velocity $\mathbf{v}_2$ can also be derived from the term

$$- q_2 \frac{\partial \mathbf{A}_1}{\partial t}$$

where $\mathbf{A}_1$ is the magnetic vector potential set up by q_1 in the place where q_2 happens to be:

$$\mathbf{A}_1 = \frac{q_1 \, v_1}{4\pi\epsilon_0 \, c^2 \, r} \, \mathbf{e}_r$$

Indeed, for charge q_2 moving at velocity $\mathbf{v}_2$ the derivative of the unit vector $\mathbf{e}_r$ is:

$$\partial \mathbf{e}_r / \partial t = \{d\theta/dt\} \, \mathbf{e}_\theta \quad \text{with:} \quad \mathbf{r} \, \{d\theta/dt\} = v_2 \text{ and:}$$

$$\frac{\partial \mathbf{A}_1}{\partial t} = \frac{v_1 \, q_1 v_2}{4\pi\epsilon_0 \, c^2 r^2} \, \mathbf{e}_\theta$$

The action of q_1 on q_2 is therefore given as follows:

$$- q_2 \frac{\partial \mathbf{A}_1}{\partial t} = - \frac{v_1 q_1 \, v_2 q_2}{4\pi\epsilon_0 \, c^2 r^2} \, \mathbf{e}_\theta$$

This force ($- q_2 \frac{\partial \mathbf{A}_1}{\partial t}$) is equal to and opposite the magnetic force $\mathbf{F}_{m21}$. However, the two forces do not have the same line of action; they are noncentral. As for the Coulomb forces of interaction, these are central in nature.

This manner of interpreting the action of charge q_1 on charge q_2—i.e., by variation of the magnetic vector potential $\mathbf{A}_1$ as seen by q_2—is confirmed by the concept of potential energy of electromagnetic interaction in this two-charge system. In fact, this potential energy E_{p12} is equal to the sum of the two terms, one electric and the other magnetic in origin:

$$E_{p12} = \frac{q_1 q_2}{4\pi\epsilon_0 \, r} - \frac{q_1 q_2 \, v_1 v_2 \, \cos\theta}{4\pi\epsilon_0 \, c^2 \, r}$$

with $\theta = \pi/2$, because the second term is obtained from the scalar product: $- q_2 \mathbf{v}_2 . \mathbf{A}_1$ (see figure 7). In mechanics, a conservative force derives from the interaction potential energy. The

67

expression of the force exerted by charge q_1 on charge q_2 is thus given by the equation:

$$\mathbf{F}_{12} = -\overline{\mathbf{V}}E_{p12} = \frac{q_1 q_2}{4\pi\epsilon_0 \; r^2} \; \mathbf{e}_r - \frac{q_1 q_2 v_1 v_2}{4\pi\epsilon_0 c^2 r^2} \; \mathbf{E}_\theta$$

In this expression, we find both Coulomb's force and the force arising from the variation of the magnetic vector potential: $-q_2 \; \partial \mathbf{A}_1$. If we accept these interpretations of the action of q_1 on q_2:

$$\mathbf{F}_{12} = \frac{q_1 q_2}{4\pi\epsilon_0 \; r^2} \; \mathbf{e}_r - q_2 \; \frac{\partial \mathbf{A}_1}{\partial t} \; .$$

and of the reaction of q_2 on q_1:

$$\mathbf{F}_{21} = -\frac{q_1 q_2}{4\pi\epsilon_0 \; r^2} \mathbf{e}_r + \mathbf{F}_{m21},$$

then there is no longer any violation of the law of equal action and reaction. Consequently, it will not be necessary to get around the difficulty by introducing (as is often done!) a certain "linear momentum" associated with these particles' electromagnetic fields in order for the law of conservation of momentum to be respected.

In any event, if the system composed of the two charges q_1 and q_2 is perfectly isolated ($\Sigma \mathbf{F}_e = 0$) in the Galilean frame of reference R_G (G being the system's center of mass), the law of conservation of momentum will be written: $\mathbf{p}_1 + \mathbf{p}_2 = 0$, with $\mathbf{p} = m\,\mathbf{w}$). We may deduce from the preceding equation that the velocities $\mathbf{w}_1$ and $\mathbf{w}_2$ of the particles are parallel and oriented in opposite directions. It will then be possible to calculate the magnetic forces exerted by the particles in question: both $\mathbf{F}_{m12}$ and $\mathbf{F}_{m21}$. It will be apparent that these forces are equal and opposite: $\mathbf{F}_{m12} = -\mathbf{F}_{m21}$. And since the Coulomb forces of interaction are themselves equal and opposite, $\mathbf{F}_{el12} = -\mathbf{F}_{el21}$, it follows that the electromagnetic forces exerted between the two moving charged particles (with respect to the reference frame of the center of mass) considered as an isolated system respect the *weak form* of Newton's third law.

Of course, if the magnetic forces do not have the same line of action, the generalized law of equal action and reaction predicts the existence of interaction couples (Γ_{12} and Γ_{21}), such that:

$$(\mathbf{M}_{12})_G = \Gamma_{12} + \mathbf{GA}_1 * \mathbf{F}_{m12}$$
$$(\mathbf{M}_{21})_G = \Gamma_{21} + \mathbf{GA}_2 * \mathbf{F}_{m21}$$

and: $(\mathbf{M}_{12})_G + (\mathbf{M}_{21})_G = 0$, with A_1 and A_2 being the positions of the charged particles q_1 and q_2. It is not unreasonable to assume that these couples do exist and that they are intimately linked to the spin-orbit or spin-spin interactions described by quantum physics.

Let us note in passing that some writers (Feynman, Purcell [8]) have made use of the interaction occurring between a rectilinear current I and a charge q_o moving at a velocity $\mathbf{v}_o$ and of the Lorentz transformation equations as a means of deducing the vectorial expression of the magnetic force exerted by the current I on the charge q_o. Such a procedure utilized a laboratory frame of reference, a frame unique to the particle q_o, and frames linked to current I's positive and negative charges. The force acting on q_o in its frame of reference is electrostatic in origin. Proceeding in this manner, and in this particular case, one can show that Newton's (generalized) third law is respected in both Galilean frames of reference considered: the laboratory frame and the particle's own frame.

Up to the present, the electromagnetic interactions that we have been considering have involved only low velocities as compared with that of light. If particles travel at velocities approaching the speed of light, it will often be necessary to describe the interaction by means of a Feynman diagram, where in addition to the two particles under consideration there will also be either a photon or a neutrino, etc. However, even under such conditions as these it is highly probable that the law of equal action and reaction in its generalized form will be respected in interactions between a body A (one of the particles) and body B (the other particle + photon + . . .). This is in keeping with

what was said in chapter 1 concerning the transformation reaction
of a proton into a neutron: $p^+ + e^- \longrightarrow n + \nu$,
i.e., proton + electron $\longrightarrow$ neutron + neutrino.

Conclusion

In chapter 3 we showed that the inertial forces that act upon
a point body are created by the interactions occurring between
two physical bodies: the material reference body moving with an
accelerated motion such as rotation and/or translation and the
point particle. We also demonstrated that these inertial forces are
similar to gravitational forces and that, like the latter, they also
respect the generalized law of equal action and reaction.

In this chapter we have examined the electrical and magnetic
interactions between various physical objects and have been led
to conclude that Newton's (generalized) third law is always respected. The example of gravitational interaction between the
sun (considered as a sphere) and the Earth (flattened at the poles)
by means of couples and noncentral forces also serves as strong
proof as to the need for generalizing the classic statement of the
law of equal action and reaction.

This generalized formulation is, moreover, in perfect accordance with experimental evidence. Indeed, interactions through
transfer of linear and angular momentum are quite general, com-
monly taking place throughout the universe in the realms of celestial mechanics, nuclear physics, chemistry, biology, etc.

Chapter 5
Newton and Relativity

In any type of physics, i.e., classical, quantum, or relativistic, describing a place where an event occurs or where an object is found means indicating the location of the reference body that coincides with the place where the event happens or where the object is located. On the surface of the Earth, a frame of reference may be created by imagining three rigid mutually perpendicular planes, fastened to a solid object. The location of an event occurring in this frame of reference is then determined by the three coordinates x, y, and z, said to be the spatial coordinates. In practice, the rigid planes are not actually drawn up, but the physical notion of our coordinates must be sufficiently precise so as to ensure that the physical phenomena observed remain distinct. Similar comments might also be made with respect to measurements of an event's "time" coordinate, which must be made by an observer bound to a given reference body. To quote Einstein:

> *Suppose that I stand at the window of a train moving at uniform speed. Suppose now that I drop a stone onto the embankment, being careful not to impart any impetus to it. What will I see? Disregarding any influence caused by air resistance, I will see the stone fall in a straight line. However, a person who happens to be outside the train will note that the stone describes a parabola as it falls. I ask now: do the various points along the stone's path to the ground really lie on a straight line or on a parabola? Here is the answer: with respect to a system of coordi-*

nates rigidly bound to the moving train car, the stone does indeed describe a straight line, but with respect to a system of coordinates bound to the ground outside, its path becomes parabolic. This example clearly illustrates that there is no absolute path per se, but only a path relative to a determined reference body [5].

In the following section we shall see how a change is made from the coordinates of an event occurring in one frame of reference to the coordinates of the same event as it takes place in another frame of reference moving with uniform rectilinear motion relative to the first frame. The speed of light is, it must be remembered, a universal constant. The Lorentz transformation equations pertaining to the coordinates (spacetime) of an event, together with equivalent transformation equations pertaining to momentum and particle energy, are the very basis of the theory of special relativity. As we shall see shortly, Newton's three laws of motion may easily be adapted to the new definitions arising from this theory and may be applied to the field of relativity. (See the sections: "The Gravitational Mass of Photons" and "The Deflection of Photons by the Sun.")

Simplified Description of Special Relativity

The experiment conducted by Michelson represents the first but not the only cornerstone in the theory of special relativity: "Any experiment involving optics or electromagnetism is proof that the theory of special relativity is sound. Both optics and electromagnetism assumed from the outset that the speed of light is constant without being aware of the immense importance that such an assumption had" (6).

The fact that the speed of light c is the same in all Galilean reference frames and that it is finite has several consequences in the field of kinematics and particle dynamics.

Kinematics

The Galilean transformation equations of an event's coordinates (x, y, z in a reference frame R; x', y', and z' in the reference frame R' moving at a speed $\mathbf{V} = V\mathbf{e_x}$ relative to R):

$$x' = x - Vt, \qquad y' = y, \qquad z' = z, \qquad t' = t \qquad (1)$$
$$x = x' + Vt', \qquad y = y', \qquad z = z', \qquad t = t'$$

leave the laws of Newtonian mechanics invariant. However, they do not leave the laws of electromagnetism invariant. If the laws of nature, including the laws of electromagnetism, are to have the same form in any system of axes in uniform rectilinear translation relative to a Galilean reference frame, then it is necessary to use the Lorentz transformation of an event's coordinates (spacetime):

$$x' = \gamma(x - Vt), \qquad y' = y, \qquad z' = z, \qquad t' = \gamma\left(t - \frac{\beta x}{c}\right) \qquad (2)$$

$$x = \gamma(x' + Vt'), \qquad y = y', \qquad z = z' \qquad t = \gamma\left(t' + \frac{\beta x'}{c}\right),$$

where $\beta = V/c$ and $\gamma = (1 - \beta^2)^{-1/2}$

These Lorentz transformation equations result in the elimination of the absolute nature of simultaneousness. Furthermore, they require that the speed of light c become a limiting speed in the universe, while modifying the theorem of addition of velocities. In fact, in classical kinematics, the theorem of addition of velocities is written:

$$\mathbf{v}^R = \mathbf{v}^{R'} + \mathbf{V}_{R'/R}$$

where R and R' are two Galilean reference frames in uniform rectilinear translation with respect to each other. If $\mathbf{V}_{R'/R} = V\mathbf{e_x}$, the formulae relating to relativistic components of velocities will be written as follows:

$$v_x = \frac{v'_x + V}{1 + Vv'_x/c^2}, \quad v_y = v'_y \frac{\sqrt{1 - V^2/c^2}}{1 + Vv'_x/c^2},$$

$$v_z = v'_z \frac{\sqrt{1 - V^2/c^2}}{1 + Vv'_x/c^2}$$

These equations restate the classical vector addition of velocities if the ratios V/c and v'_x/c are much less than 1.

We already know that these formulae have been verified hundreds of times through experimentation. Let us briefly mention the experiment of Fizeau, who in 1851 investigated the speed of light in a stream of water. In an effort to simplify the calculations, we shall assume that the liquid is benzene through which light is found to travel at the speed of 200,000 kilometers per second. If the benzene now flows through a tube at the speed of 50 km./s., classical mechanics predicts that the ray of light should acquire a speed of 200,050 km./sec with respect to the observer as it is carried by the moving liquid. Actual measurements reveal a speed of 200,028 km./sec, a value confirmed by the formula of relativistic addition of velocity:

v'_x = 200,000 km./s., V = 50 km./s., c = 300,000 km./s.,

$$v'_y = v'_z = 0$$

Relativistic Dynamics

Einstein noticed that there was a complete similarity between an "event" and a "particle" and that $(p_x, _y, _z; E/c)$ could be made to correspond to (x, y, z, ct), where p_x, p_y and p_z are the components of the particle's momentum and E is its energy. This property enables us to write directly the Lorentz transformation equations for momentum and energy. All one has to do, in fact, is replace x b y p_x, y by p_y, z by p_z and ct by E/c in equation (2) above. The same applies to the primed variables. We thus obtain:

$$P'_x = \gamma (p_x - \beta \frac{E}{c}) , p'_y = p_y, p'_z = p_z, E' = \gamma (E - \beta c p_x),$$

$$P_x = \gamma (p'_x + \beta \frac{E'}{c}), p_y = p'_y, p_z = p'_z, E = \gamma (E + \beta c p'_x),$$

where β and γ have the same meanings as above in (2).

As for the universe interval Δs^2, which is independent of reference frames R or R' in which it is evaluated:

$$\Delta s^2 = d^2 - c^2 \Delta t^2 = d'^2 - c^2 \Delta t'^2 = \Delta s'^2$$

the equation of invariance pertaining to the particle will be written:

$$m^2 c^2 = (E/c)^2 - p^2 = (E'/c)^2 - p'^2$$

where m is, by definition, the particle's constant mass (corresponding to its mass at rest in textbooks where mass is assumed to change with the particle's velocity).

In light of the preceding equations, we may easily derive new definitions of the momentum **p** and relativistic energy E of a particle having a mass m as evaluated with respect to a reference frame R. We obtain:

$$\mathbf{p} = \frac{m\,\mathbf{v}}{\sqrt{1 - v^2/c^2}} \quad \text{and:} \quad E = \frac{m\,c^2}{\sqrt{1 - v^2/c^2}}.$$

Some comments are in order here. First, we would do well to point out that in the formulae defining the relativistic momentum **p** and energy E, it is the velocity **v** of the particle relative to R that appears, whereas in coefficients β and γ of the Lorentz transformation equations it is the velocity $\mathbf{V} = V\mathbf{e}_x$ of reference frame R' relative to reference frame R that is involved.

Second, with respect to particles having zero mass (m = 0): the invariance equation $m^2\,c^2$ in the Lorentz transformation pertaining to momentum and energy leads one to imagine the existence of massless particles (m = 0). These particles, which are called "photons," move at the speed of light c with respect to any Galilean reference frame. In addition, they convey momentum and relativistic energy as expressed in the following equations:

$$p = h\nu/c \text{ and } E = h\nu,$$

where ν is the frequency of the associated electromagnetic wave and $h = 6.624 \cdot 10^{-34}$ J.s is Planck's constant. Besides photons, there are other such particles: neutrinos, antineutrinos, etc.

Newtonian Laws and Special Relativity

Newton's first law, or law of inertia, remains valid in special relativity. The definitions of the Copernican and Galilean reference frames are derived from it. Special relativity, on the other hand, is based on the following postulate: "All laws of physics must have the same form in any Galilean reference frame and must be invariant by the Lorentz transformation."

Such is the case with Newton's first law, even if we consider the generalization of this law (see chapter 1) which deals with uniform rotational motion of various studied objects.

It is also the case with Newton's second law (at least when it applies to a particle) if we take into account the definition of the particle's relativistic momentum. Indeed, in a Galilean reference frame R, the basic principle of mechanics as applied to a particle having a mass m will be written thus:

$$\mathbf{F} = m d \left[\mathbf{v}(1 - v^2/c^2)^{-1/2}\right]/dt = d\mathbf{p}/dt$$

Generally speaking, force is no longer proportional to acceleration. This is due to the factor of relativistic correction which is involved in the derivative of the vector $\mathbf{v}(1 - v^2/c^2)^{-1/2}$. It should be said that force, mass or acceleration can only be introduced in special relativity with caution and, in particular, only when the material systems studied are simple and well defined.

In a Galilean reference frame R' moving with a velocity $\mathbf{V} = V\mathbf{e}_x$ relative to the preceding reference frame R, the basic principle applied to the same particle with mass m will be written:

$$\mathbf{F}' = m d \left[\mathbf{v}'(1 - v'^2/c^2)^{-1/2}\right]/dt = d\mathbf{p}'/dt$$

All primed quantities are evaluated with respect to R'. Forces no longer have expressions independent of the reference frames, as they do in classical mechanics (v, v', V << c). We can establish the transformation equations pertaining to the forces by using the Lorentz transformation equations for momentum, Newton's second law and the theorem of kinetic energy.

Theorem of Kinetic Energy

In special relativity, this theorem can be established very easily starting from classical definitions of the elementary work involving a force $\mathbf{F}$ (dW = $\mathbf{F}.\mathbf{dr}$, where $\mathbf{dr}$ is the elementary displacement of the application point of this force as evaluated in a frame of reference R) and the relativistic energy of a particle having a mass m. In fact, the power of the force $\mathbf{F}$ acting upon the particle is written:

$$\frac{dW}{dt} = \mathbf{F} . \mathbf{v} = \mathbf{v} . \left(\frac{m\,\mathbf{v}}{\sqrt{1 - v^2/c^2}} \right)$$

After several simple calculations we obtain:

$$\frac{dW}{dt} = \frac{m\,v\,(dv/dt)}{(1 - v^2/c^2)^{-3/2}}$$

The time derivative of the particle's relativistic energy is given by:

$$\frac{dE}{dt} = \frac{d}{dt} \frac{m\,c^2}{\sqrt{1 - v^2/c^2}} = \frac{m\,v\,(dv/dt)}{(1 - v^2/c^2)^{3/2}}$$

From these calculations it is obvious that dW/dt = dE/dt and, since relativistic energy is equal to the sum of the particle's relativistic kinetic energy and of its own energy as expressed by $E_0 = mc^2$, $(E = E_{cr} + E_0)$, we can write:
dW/dt = dE/dt = dE_{cr}/dt, and, for a finite interval of time Δt:
$\Delta E = \Delta E_{cr} = \Delta W_{(F)}$.

"The variation of relativistic kinetic energy (or of relativistic energy) is equal to the work of Force $\mathbf{F}$ acting upon the particle between the initial position and the final position." This statement is a generalization of the theorem of kinetic energy as applied to a particle in classical mechanics.

Newton's Third Law

The interpretation of the interactions occurring between particles in the area of relativity (with the speed of particles approaching that of light) is, as indicated at the end of chapter 4, rather delicate. If we limit ourselves to an isolated system of two particles, we may apply in a Galilean reference frame successively to two particles:

$$d\mathbf{p}_1/dt = \mathbf{F}_{21} \quad \text{and} \quad d\mathbf{p}_2/dt = \mathbf{F}_{12}$$

where $\mathbf{F}_{12}$ is the force exerted by particle 1 on particle 2 and $\mathbf{F}_{21}$ is the force exerted by 2 on 1. $\mathbf{p}_1$ and $\mathbf{p}_2$ are the relativistic momenta of the two particles. If we accept as a basic postulate that there must be conservation of total relativistic momentum, then we shall have:

$$\mathbf{p}_1 + \mathbf{p}_2 = \text{constant}, \quad \frac{d\mathbf{p}_1}{dt} + \frac{d\mathbf{p}_2}{dt} = 0 \text{ and } \mathbf{F}_{21} + \mathbf{F}_{12} = 0$$

The last equation represents one of the two conditions of the law of equal action and reaction. It is quite probable that the second condition pertaining to resulting moments is also respected since we accept the other two laws of conservation of total angular momentum (including spin angular momentum) and relativistic energy in an isolated system as postulates. (See the discussion on Feynman diagrams on page 69.)

The Mass–Energy Equivalence

The law of conservation of relativistic energy in an isolated system of particles is not based on the assumption of constancy (i.e., conservation) of the system's total mass, as was the case in Newtonian mechanics. Consider the ideal example of a perfectly inelastic head-on collision between two identical particles (m) yielding only one particle (m_3) at rest in the reference frame R_G of the center of mass. The equation expressing the conservation of relativistic energy in the isolated system ($E_1 + E_2 = E_3$) will be written as follows:

$$\frac{mc^2}{\sqrt{1 - v^2/c^2}} + \frac{mc^2}{\sqrt{1 - v^2/c^2}} = m_3c^2$$

78

From this equation is drawn $m_3 = 2m /\sqrt{1 - v^2/c^2}$, where v is the common modulus of the speeds of the incident particles (before collision).

We note that the mass (m_3) of the particle at rest in R_G is greater than the sum of the masses of the incident particles $(m_3 > 2m)$. The system's total relativistic kinetic energy, which is null following the collision, was prior to collision equal to:

$$E_{cr} = 2m\,c^2\,[(1 - v^2/c^2)^{-1/2} - 1]$$

If we write $m_3 = 2m + \delta m$, where δm is the excess mass in this particular instance, we can write $E_{cr} = c^2\,\delta m$. It follows then that the kinetic energy that was available prior to the collision has been "transformed" following collision into mass $(m_3 > 2m)$.

In certain reactions between particles in an isolated system (fission and fusion, for instance), it sometimes happens that the system's final total mass is less than initial total mass. In this case, the missing mass δm is found in the form of energy radiated by the system.

Generally speaking, any variation of energy δE (kinetic, potential, etc.) in an isolated system of particles is due to a variation in the system's total mass as expressed in the following equation:

$$\delta E = c^2\,\delta m$$

Absorption of energy by the system of particles corresponds to the increase in the system's mass. Energy radiated by the system of particles corresponds to a decrease in the system's mass.

Einstein considered this equivalence relation between mass and energy to be the most striking result of the theory of special relativity. Indeed, thermonuclear energy provides an interesting practical application of this formula. Reactions involving the fusion of light nuclei (hydrogen and its isotopes) are the main source of energy radiated by the sun and the other stars. Such fusion reactions take place easily because the nuclei have extremely high kinetic energies (the temperature at the center of the sun being 2.10^7 °K). Although these reactions actually occur in several stages, the following simplified description illustrates the basic process:

4 protons + 2 electrons ---▶ 1 He^4 (or α particle) + δE

Since we know that the masses of a proton, an electron, and α particle are respectively $1.6725.10^{-27}$, $0.0009.10^{-27}$ and $6.647.10^{-27}$ kg., we can calculate the missing mass and the amount of energy released in the fusion reaction: δE = c^2 δm = 25 Mev.

Unfortunately, technological limitations being what they are, it has not yet been possible to build a self-sustaining thermonuclear reactor, i.e., one in which the energy released by fusion might serve to set off other reactions while yielding substantial amounts of energy with no danger of exploding.

The Gravitational Mass of Photons

As we have seen previously, a photon moves at the speed of light (c) relative to any Galilean reference frame, conveying both momentum (hν/c) and energy (hν). Since the momentum is equal to the product of inertial mass and the speed of the particle in Newtonian mechanics, we thus infer that the photon has an "inertial mass" equal to hν/c^2. It must be remembered that the mass of a photon at rest is null.

According to Einstein's equivalence principle, the photon should have a "gravitational mass" also equal to hν/c^2. This property was revealed in an experiment carried out at Harvard by two American physicists (R. V. Pound and G. A. Rebka, *Phys. Rev. Letters*, 4, no. 337 [1960]).

In a simplified description of this experiment, let us consider a photon with a frequency ν and having an energy hν. Suppose now that it has been emitted by a source located at a given height H above the Earth's surface. The energy of the photon increases by mgH (m = hν/c^2, g = weight acceleration) when it falls from height H. The law of conservation of energy requires that the photon's energy $h\nu'$ upon arrival on the ground be:

$$h\nu' = h\nu + \frac{h\nu}{c^2} gH$$

This is true if we assume that the gravitational mass of the photon remains constant during its fall, a reasonable assumption since $h\nu'$ is very nearly equal to $h\nu$. The receiver placed on the ground will detect a frequency slightly greater than the frequency ν at the source:

$$\nu' = \nu \left(1 + \frac{gH}{c^2}\right).$$

This effect of frequency shift is very weak as measured in the experiment conducted by Pound and Rebka: $H = 22m$, $\frac{\Delta\nu}{\nu} = 2.4.10^{-15}$. However, it may become relatively important in the case of photons emitted by a star whose mass M_e is close to that of the sun and whose volume approaches that of the Earth. Such a star is O_2 Eridani B, a white dwarf having an extremely powerful gravitational field on its surface. Spectral shifts that have confirmed the results of the theory have been detected

$$\nu' = \nu \left(1 - \frac{GM_e}{R_e c^2}\right) \text{ with}$$

$$\frac{\Delta\nu}{\nu} = -6.10^{-5}$$

where ν is the frequency of the photons emitted by the star's surface, ν' is the frequency detected at the Earth's surface, G is the universal constant of gravitation, and R_e is the radius of the star in question.

The Deflection of Photons by the Sun

Light passing near a star must be deflected, as any material projectile would be if it were subjected to the star's gravitational attraction. The problem involves photons moving at the speed of light in the gravitational field. Suppose that these photons have a gravitational mass m. Suppose also that they describe a rectilinear path as if they underwent very little deflection and that they pass at a minimum approach distance r from the center of the sun (a distance slightly greater than the radius of the sun R_s. In figure 8, two straight lines are shown: D and D'. The first line, D, would be the undeflected path of photons emitted by a remote star parallel to direction Oy. The second line D', which makes a very small angle α with D, represents the "deflected" path of the photons. α is so small that P and P' may be considered as practically one and the same (OP = OP' = r). The transversal component F_x representing the force of gravitational attraction acting on the photon, located at the point as defined by the coordi-

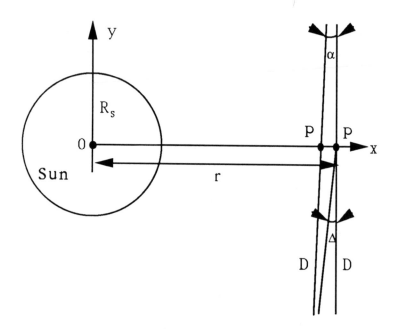

Figure 8. The Deflection of Photons by the Sun

nates r and y, is given by the formula:

$$F_x = -\frac{GM_s mr}{(r^2 + y^2)^{3/2}} \ ,$$

where y is measured from point P shown in figure 8.

The final value of the transversal component v_x of the photon's speed is: $mv_x = \int F_x dt = \int F_x (dy/v_y) = \frac{1}{c} \int F_x dy$

whence:

$$v_x = -\frac{2GM_s r}{c} \int_0^\infty \frac{dy}{(r^2 + y^2)^{3/2}} = -\frac{2GM_s}{c \ r}$$

The angle of deflection, α, or the angle formed by the two straight lines D and D' is given by the equation: $tg\alpha = \alpha = v_x/c$, and when r approaches the radius of the sun R_s, we have: $\alpha = \frac{v_x}{c} = \frac{2GM_s}{c^2 R_s}$. Finally, if the angle of deflection is measured from point P in minimum approach to the sun, we will have to consider the angle:

$$\Delta = 2\alpha = \frac{4GM_s}{c^2 R_s} \ ,$$

which corresponds to the value as predicted by the theory of relativity.

Use of the Theory of Relativity to Calculate the Perihelion Advance of Certain Planets

The theory of relativity has one consequence that is of particular interest to celestial mechanics. This is the law of gravity of an infinitely small mass under the influence of fixed masses. In its formulation, this law of gravity is equivalent to Newton's law at first glance, but in fact represents a correction to Newtonian law. As a theory, it is more powerful in explaining a noteworthy perturbation that did not conform to Newtonian mechanics: the advance of perihelion of the planet Mercury.

83

In the case of planets subjected to the sun's gravitational attraction, this law of gravity may be stated as follows: "under the influence of a spherical mass which is isolated and not rotating, the motions of a small mass m will be defined by the geodesics of Schwarzschild's ds^2":

$$ds^2 = (c^2 - \frac{2GM_s}{r}) dt^2 - \frac{dr^2}{(1 - \frac{2GM_s}{c^2 r})} - r^2 (d\theta^2 + \cos^2\theta\, d\phi^2)$$

where c is the speed of light, G the universal constant of gravitation, M_s the mass of the sun, r ϕ the spherical polar coordinates of the planet in a Copernican reference frame, and t the absolute cosmic time.

A presentation of the methods used to resolve differential equations for motions inferred from this ds^2 goes beyond the scope of this book. Readers interested in such a discussion should consult J. Chazy's excellent work (7). We shall therefore limit ourselves here to the most important results concerning a planet's motion around the sun.

First of all, such motion occurs in the plane determined by the center of the sun and the planet's initial velocity. As Kepler's law relating to elliptical motion explains, the planet follows an orbit that is an ellipse. However, an important difference now appears. Instead of remaining fixed in space, the planet's major axis slowly shifts in space, turning at a very small angle $\Delta\theta$ during one revolution around the sun. This perturbation is defined as the advance of perihelion, and the formula that expresses this advance at each revolution may be given as follows:

$$\Delta\theta = \frac{6\pi GM_s}{c^2 a (1 - e^2)}$$

In this expression a represents half of the major axis of the ellipse and e is its eccentricity.

If we assume the motions of the planets in the solar system to be circular, with a being the radius of the circle, e = 0, $GM_s/r = v^2$ where v is the planet's velocity relative to the Copernican reference frame centered on the sun, then we will obtain a formula that is naturally only approximate, but which nevertheless yields numerical values very close to those in the former equation:

$$\Delta\theta = 6\pi\, v^2/c^2$$

This formula reveals the relativistic nature (factor $3v^2/c^2$) of the correction to Newton's law ($\theta = 2\pi$ for one revolution).

To sum up then, according to Newton's law and in agreement with Kepler's first law, the planets in their revolutions around the sun describe ellipses fixed in their planes. Einstein's law states that these ellipses slowly turn in their planes. Consequently, account must be taken of a shift in the perihelion of planetary orbits.

Calculating Advance of Perihelion Using Newton's Second Law

The study of the similar behavior of all forces of interaction (between physical objects) in terms of Newton's three laws of motion leads us to make an assumption that further tends to unify the behavior of electromagnetic and gravitational forces. To illustrate this, suppose that the forces of gravitational attraction exerted by the sun on its planets have a "relativistic" component in the same way that the force exerted by a charged particle q_1 (moving at velocity v_1) on a second charge particle q_2 (moving at velocity v_2) has a "relativistic" component—the magnetic force $q_2 v_2 * B_1$, with $B_1 = \dfrac{v_1 * E_1}{c^2}$. Then a "gravirotational" field

h would have to exist at the place where the planet (assumed to be a point) is located, in such a way that the force exerted by the mass M_s on the mass m_p would be given by: $F_{sun-planet} = m_p g + m_p v_2 * h$, with: $g = \dfrac{-GM_s}{r^2} e_r$, $h = \dfrac{v_1 * g}{c^2}$, and $v_1 = -\Omega_s * r$, where Ω_s is the angular velocity of the sun's rotation, $\mathbf{r}$ is the position-vector of the mass m_p relative to the center of the sun, and $\mathbf{g}$ is the gravitational acceleration due to mass M_s at the place where m_p is found.

We know that the paths of the planets around the sun are ellipses located in planes close to that of the sun's equator (in the case of Mercury, these planes are one and the same). The "gravirotational" field $\mathbf{h}$ due to the sun's rotation on its axis $\mathbf{e}_z$ is thus given by the equation:

$$\mathbf{h} = \frac{-GM_s\Omega_s \, \mathbf{e}_z}{rc^2}.$$

Like its analogue the magnetic field $\mathbf{B}$, the field $\mathbf{h}$ derives from a vector potential:

$$\mathbf{A} = \frac{-GM_s\Omega_s \, \mathbf{e}_\phi}{c^2} \quad (\mathbf{h} = \mathrm{rot} \, \mathbf{A})$$

where $(\mathbf{e}_r, \mathbf{e}_\phi, \mathbf{e}_z)$ are the bases of a system of cylindrical polar coordinates and the Lagrangian of the system consisting of the planet is:

$$L = \tfrac{1}{2}m_p v^2 + \frac{GM_s m_p}{r} + m_p \mathbf{v} \cdot \mathbf{A}, \text{ meaning:}$$

$$L = \tfrac{1}{2}m_p \left\{ \left(\frac{d}{dt}\right)^2 + r^2\left(\frac{d\phi}{dt}\right)^2 \right\} + \frac{GM_s m}{r} - \frac{GM_s m_p \Omega_s}{c^2} \, r \, \frac{d\phi}{dt}$$

Lagrange's equation associated with the generalized coordinate ϕ is a first integral of the motion:

$$\frac{\partial L}{\partial \phi} = m_p r^2 \frac{d\phi}{dt} - \frac{GM_s m_p \Omega_s \, r}{c^2} = p\phi = \text{constant}.$$

Let us divide by m_p and assume that the motion of the planet is circular with a period T. We may then write the preceding equation in the form:

$$\frac{2\pi r^2}{T} - \frac{GM_s\Omega_s r}{c^2} = \frac{2\pi r_0^2}{T_0}$$

where T_0 and r_0 represent respectively the period and the radius of the circular path if we disregard the force $m_p \, \mathbf{v} * \mathbf{h}$, and where the second term on the left is a "relativistic" correction factor.

As r is nearly equal to r_0 and $T_0 - T = \Delta T$ with T nearly equal to T_0, we may deduce from the last equation the expression of relative decrease in the period caused by the "relativistic" component (m_p **v** * **h**) of the sun's gravitational force of attraction:

$$\Delta T/T = \frac{GM_s \Omega_s}{2\pi c^2} \frac{T}{r}$$

This relative decrease in the period results in an advance of the planet's perihelion.

This formula may be applied to the planets in our solar system as well as to artificial satellites orbiting the Earth (provided that they orbit in a counter-clockwise direction) in the Earth's equatorial plane; here it is necessary to replace Ω_s with Ω_e and M_s with M_e. If, on the other hand, the artificial satellites orbit the Earth in the direction opposite that of the Earth's rotation, there will be a relative increase in period, meaning a delay of perihelion.

The Secular Perihelion Advance of the Planets: Theories and Experimental Evidence

1. A Review of the Formulae

a) General relativity. In light of what we stated above, the secular advance δ of the planets according to the theory of general relativity would be:

$$\delta = n\Delta\theta = \frac{6\pi GM_s n}{c^2 a (1 - e^2)}$$

where n is the number of revolutions completed by the planet during one hundred years (n = 36525/T, if the period T is expressed in days).

b) "Relativistic" gravitational attraction. The formula as drawn from Newton's "relativistic" laws (analogous to the laws of electromagnetism) predicts a relative decrease in the period of revolution for planets in the solar system rotating in a counterclockwise direction. The equation will be:

$$\Delta T/T = \frac{GM_s \Omega_s}{2\pi c^2} \frac{T}{r}$$

where for the average value of Ω_s we will choose $2.42 \ 10^{-6}$ rad/s. This corresponds to one complete rotation of the sun in thirty days. As a result, the secular advance δ' of the perihelion of the planets according to this theory will be:

$$\delta' = n \left(\frac{\Delta T}{T}\right) 2\pi$$

2. Possible Comparisons

In order to compare the theoretical values calculated from the two formulae given above with actual results obtained from astronomical observation, we provide below the values drawn from Newcomb's Table as it appears in Chazy's work (7), p. 171. This table is a summary of astronomical observations made by Newcomb in 1895 of the advances of perihelion of the four small planets found in the solar system.

Table I

Secular advances in the perihelions of the four small planets				
Value given by Le Verrier in 1859 for Mercury: 39"				
Secular advance	Mercury	Venus	Earth	Mars
Exp. (sec. of arc)	$41.24 \pm 2''$	$-8'' \pm 37''$	$6'' \pm 8''$	$8'' \pm 4''$
Theor. δ	42.9"	8.61"	3.83"	1.35"
Theor. δ'	40.5"	21.6"	15.5"	10.2"

The numerical values provided by the two formulae δ and δ' agree with the experimental results with respect to the advance of perihelion of Mercury. The second (δ') reveals more clearly that of Mars. The validity of this new formula could be confirmed (or invalidated) by precise investigations made on the motion of asteroids (minor planets) around the sun. In Table II we have indicated the calculated values δ and δ' of secular advances for some of these asteroids. It would be interesting to compare these with observed values if they exist.

Table II

Secular advances in the perihelions of a few minor planets					
Minor Planet	Mean distance from Sun r	Semi-major axis a	Eccentricity e	Sidereal period (T)	δ	δ'
units	10^6km	10^6km		days	sec. of arc	
Icarus	200	161.5	0.83	408.8	10.3″	11″
Apollo	250	220	0.56	650	2.1″	8.8″
Adonis	350	280	0.764	935	1.9″	6.2″
Ceres Pallas	414	414	0.165	1680 1684	0.31″	5.3″

The Dilation of Time

Let there be two clocks: H and H', one fixed and the other (H') moving at velocity v with respect to the first clock. According to the theory of special relativity, clock H' should run slow relative to clock H. This phenomenon is known as the dilation of time.

Let there now be two clocks H and H', one (H) on the Earth's surface and the other (H') at a given altitude h. According to the theory of general relativity, clock H', subjected to less gravitational attraction, should run fast with respect to clock H.

Two American scientists, J. C. Hafele and R. E. Keating (*Science*, 177 [1972]:166) actually verified these predictions in an experiment in which they sent two atomic clocks H'_+ and H'_- around the Earth in Boeing 747s. The times shown by these two clocks at the beginning and end of the trip were compared to the reference clock H located in Washington. Clock H'_+ made one complete revolution around the Earth moving from west to east, while H'_- made the trip from east to west.

The experimental results, in agreement with the two theories outlined above, showed that H'_+ was slightly behind and H'_- slightly ahead of the reference clock H. These results bear out the validity of Einstein's theories, but they also prove Newton's theory inasmuch as the interpretation of these results must hinge on the use of an "absolute" reference frame, whose origin is the center of the Earth and whose axes remain parallel to the Copernican frame of reference.

Conclusion

If Newton's laws borrow the Lorentz transformation equations and the new definitions of a particle's momentum and energy from the theory of special relativity, it will be found that they remain applicable in the relativistic area. In addition, they can explain the deflection of light by the sun as well as the shift toward the red in radiation emitted by a surface of a star.

With respect to the secular perihelion advances of the planets in the solar system, the situation is not as simple, a fact shown by the number of theories published on this subject. Those who defend the theory of general relativity tend to look at only the advance of the perihelion of Mercury, commenting, like Einstein, that "for the other planets in our solar system this advance is so small that it cannot be detected" (5).

The formula that I propose does not claim to be able to solve the problem in a definitive manner. However, as it does provide theoretical values close to the experimental values for the secular advances of the first four planets of our solar system, I think it is interesting to submit the formula to astronomers and other physicists who are concerned with the motion of artificial satellites orbiting the Earth as well as the asteroids orbiting the sun.

Chapter 6
Some Applications

I. Theorem of Angular Momentum Applicable to a Variable-Mass Solid

Let us consider as an example a circular disk that can revolve around a vertical axis with no friction. Standing on this disk is a man, motionless, holding at arm's length two identical masses $m_1 = m_2 = m$. Suppose now that the material system made up of the disk, the man, and the two masses is left to itself once it has been set in motion. It is thus "isolated," because the sum of external forces acting upon the system is null. According to the theorem of angular momentum we have:

$(I_\Delta + I_h + 2mr^2)(d\omega/dt) = 0$, whence: $\omega = \omega_0$,

where I_Δ, I_h and $+ 2mr^2$ are the moments of inertia of the disk, the man, and the two masses relative to the vertical axis of rotation.

At time t, the man is asked to drop the two masses m_1 and m_2. According to the theorem of angular momentum applied to the system that has just lost the two masses we should write:

$(I_\Delta + I_h)(d\omega/dt) = M_{O\ (F\ ext.)}$

where $F_{ext.}$ is reduced to the term $\mathbf{F}_{interaction}$, forces possibly exerted by the masses m_1 and m_2 on the man at the moment they are separated from the initial system. However, this term is null inasmuch as the masses m_1 and m_2 leave the system with a relative velocity that is null. As a result, the second member of the preceding equation is null: $M_{O\ (F\ ext.)} = 0$ and we must have $d\omega/dt = 0$ as well, whence: $\omega = \omega_O = $ constant.

The angular momentum of the initial system (disk + man + two masses m_1 and m_2) was $(I_\Delta + I_h + 2mr^2)\ \omega_O$ before the two masses were dropped, and the angular momentum of the system having lost these two masses has become $(I_\Delta + I_h)\ \omega_O$. The angular momentum of the overall system (revolving disk + man + the two masses), which we may consider as isolated, is nevertheless actually conserved since to the term $(I_\Delta + I_h)\ \omega_O$ must now be added the angular momentum of the two masses m_1 and m_2 with respect to the vertical axis of rotation. It may be easily calculated: $2mr^2\ \omega_O$.

This example is of interest because it can be verified readily through experimentation (which is what we have done) and thus contributes to the proof that the theorem of angular momentum when applied to a variable-mass solid should be written:

$$[I_O(t)\ d\omega/dt] = M_O\ (F_{interactions} + F_{autres})$$

This equation may be compared to equation (2) on page 20.

Although lawn sprinklers are not "variable-mass solids," they are often used to illustrate the force of interaction exerted by the gushing water on the sprinkler's arms. Writers who make use of Newton's third law in determining the motor couple are indeed few and far between. Still more exceptional are those who mention experimental results. Let us mention that our results (J. M. Rocard and B. Marais, "Tourniquets hydrauliques," *La houille blanche* 6 [1989]: 420) are in complete disagreement with the officially accepted theoretical formula.

It seems that the interaction through transfer of momentum is the same in the examples of the lawn sprinkler and the rocket. The expression of the force of interaction should be the same in both instances. It goes without saying that there are differences between the rocket and the sprinkler, and that no attempt is being made to hide this fact. The rocket carries its source of energy with it, whereas the lawn sprinkler is connected to its energy source. A "braking" couple is involved when water moves in the sprinkler's arms, but its expression should depend on the form of the lawn sprinkler under consideration, which is not the case in the officially accepted formula.

II. The Dipole-Point Charge Electrostatic Interaction

Figure 9 shows an electric dipole placed at the origin O of the reference frame R and a point charge $+q$ located at a point (B) far from the dipole. We note that in the particular circumstances shown in figure 9 the mutual actions of both bodies (dipole p = body A, point charge q = body B) involve noncentral forces of interaction as well as a couple.

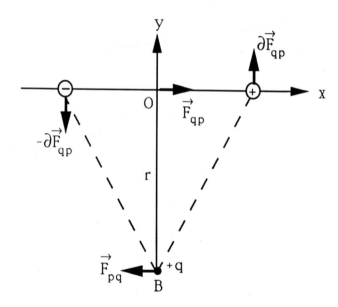

Figure 9. The Dipole-Point Charge Electrostatic Interaction

94

In fact, the action of body A on body B is represented by force $\mathbf{F}_{pq}$ which is parallel to the dipole and applied at B. The action (or reaction) of body B (the charge q) on body A (dipole p) consists of a noncentral force $\mathbf{F}_{qp}$ applied at the center O of the dipole and a couple with axis Oz: Γ_{qp} (in figure 9 represented by the two forces $\pm \partial\mathbf{F}_{qp}$).

In accordance with the generalized law of equal action and reaction we must write:

$\mathbf{F}_{pq} + \mathbf{F}_{qp} = 0$ and $(\mathbf{M}_{pq})_0 + (\mathbf{M}_{qp})_0 = 0$

where the resulting moment at O of the action of q on p is $(\mathbf{M}_{qp})_0 = \Gamma_{qp}$ and that of the action of dipole p on charge q is: $(\mathbf{M}_{pq})_0 = \mathbf{OB} * \mathbf{F}_{pq}$. We note that OB $= r$ and suppose that $r >> d$, where d is the distance separating the two charges of the dipole. Forces $\mathbf{F}_{pq}$, $\mathbf{F}_{qp}$ and couple Γ_{qp} are calculated from the electrostatic laws:

$$\mathbf{F}_{pq} = q\,\mathbf{E}_\theta = -\frac{p\,q}{4\pi\epsilon_0 r^3}\,\mathbf{e}_x,$$

where $\mathbf{E}_\theta$ is the electric field created by the dipole at B. The force:

$$\mathbf{F}_{qp} = +\frac{p\,q}{4\pi\epsilon_0 r^3}\,\mathbf{e}_x, \text{ is a projection onto the x axis of the}$$

electrostatic forces of interaction of charge q with each of the dipole charges. The couple exerted by a charge q on dipole p is:

$$\Gamma_{qp} = \mathbf{p} * \mathbf{E} = \frac{p\,q}{4\pi\epsilon_0 r^2}\,\mathbf{e}_z,$$

where $\mathbf{E}$ is the electric field created by charge q at 0. And finally,

$$\mathbf{OB} * \mathbf{F}_{pq} = -\frac{p\,q}{4\pi\epsilon_0 r^2}\,\mathbf{e}_z$$

From these expressions it is simple matter to infer:

$\mathbf{F}_{pq} + \mathbf{F}_{qp} = 0$ and $\Gamma_{qp} + \mathbf{OG} * \mathbf{F}_{pq} = 0$, meaning:

$$(\mathbf{M}_{qp})_0 + (\mathbf{M}_{pq})_0 = 0$$

We could demonstrate that this result is general. Regardless of the position of the point charge $+q$ relative to the dipole, the mutual actions of both bodies (dipole-charge) actually respect

95

the generalized law of equal action and reaction. Naturally, the moments resulting from these actions are equal and opposite only when they are evaluated at the same point in space.

III. Theorem of Energy Applicable to the Rocket

There are two ways to apply the law of conservation of energy in the problem involving the rocket (p. 22) moving in a Galilean frame of reference in the absence of any gravitational field. We consider the material system with constant total mass as composed of the rocket, the fuel, and the combustive. Energy totals are then determined either in the non-Galilean frame of reference bound to the rocket or in the Galilean frame with respect to which the rocket is moving.

In the non-Galilean frame of reference, the energy available at the time of departure is the chemical energy E_0 of the fuel and the combustive carried off by the rocket. In this frame, the chemical energy is transformed completely into relative kinetic energy of the exhaust gases $(1/2)m_0u^2$ (u = the relative speed at which the gases are ejected; m_0 = the initial mass of both fuel and combustive).

In the second (Galilean) reference frame, the same chemical energy is transformed into final kinetic energy of the space vehicle $(1/2)\,m_cv_f^2$ and into absolute kinetic energy (E_{cg}) of the exhaust gases. Calculating the integrals, which we shall not reproduce here owing to space limitations, shows that we indeed have the following:

$$E_0 = \tfrac{1}{2}m_0u^2 = \tfrac{1}{2}m_cv_f^2 + E_{cg}$$

(For more details, see J. M. Rocard, "Théorème de l'énergie appliqué à la fusée," *European Journal of Physics* 11 [1990]: 308–10.)

IV. The Secular Slowing Down of the Earth as It Rotates on Its Polar Axis

It has been known for a long time that the length of a day, equal to the 24-hour period of the Earth's rotation on its axis, has increased by one or two seconds in the last 120,000 years. This slowing down is caused by frictional forces created as the oceans move over the earth's crust. In other words, the phenomenon can be explained by taking into account the tides occurring on the Earth's surface.

The purpose of this section is to show that this slowing down is a clear indication of the generalized law of equal action and reaction.

As is the case in most classical astronomy texts, we shall resort to a few assumptions in an effort to make it easier to understand the phenomena we are dealing with. We shall assume first of all that the plane of the ecliptic (the path of the center O of the Earth around the sun) and the plane of the path of the center of the moon around the Earth are one and the same. In fact, the angle that these two planes make is very small, equal to 5°9'. We shall next assume that the Earth's equatorial plane is the same as that of the ecliptic. This assumption stretches reality a bit, since the angle made by the two planes in question is 23°26', but does not invalidate the reasoning that we are now going to follow. Finally, we shall assume that the sun and the moon are in conjunction or opposition, which is the case at times of the greatest tides. The phenomena are similar when the sun and the moon are in quadrature, although less noticeable. In figure 10 we have shown in the plane of the ecliptic how the Earth's equatorial circle, the moon and the two tidal bulges (D_1 and D_2) look at the time of year when tides are greatest.

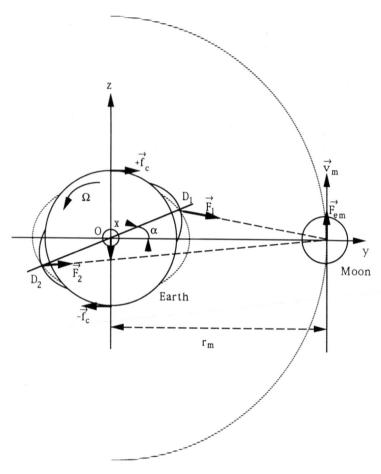

Figure 10. Diagram showing the relative Earth–moon arrangement and bulges in the surfaces of the oceans when the sun and moon are in opposition.

Tidal Effects

The effect of tides on the surface of the Earth (which is seven-tenths oceans and three-tenths continents) is due to driving

98

inertial forces acting on the water masses in the oceans, because the motions of the Earth around the sun and of the moon around the Earth are practically uniform and circular, hence accelerated. The two bulges (D_1 and D_2) and the period (24 hours) of the Earth's rotation on its polar axis explain why there are about two high tides and two low tides daily at any given location on the Earth. If there were no friction holding back water on the Earth's crust, the bulges in the oceans' surface should be found directly opposite the moon on the Oy axis, as well as in the area diametrically opposite, on the other side of the Earth, as shown by the dotted lines in figure 10. However, these frictional forces do exist, in addition to the forces of intramolecular cohesion that hold water molecules close to each other. These forces cannot be disregarded. Since the angular velocity of the water masses located at D_1 and D_2 is 29.5 times greater than that of the lunar vector radius, the driving effect imparts to direction D_1 D_2 an angular rotation α on the axis Ox (see figure 10). This factor of 29.5 is linked to the value of the synodic revolution, which is the interval of time separating two consecutive conjunctions of the moon with the sun, i.e., two consecutive new moons (on average 29 days, 12 hours, 44 minutes).

Assessing Forces

Let us now assess the forces acting on the Earth–moon system under the conditions shown in the figure. We isolate this system so as not to have to reckon with the forces of attraction exerted by the sun on the Earth and on the moon. Of course they are involved, since some of them contribute to the tidal effects that we are discussing. However, they lie outside the system. What we are attempting to describe are the forces of interaction between the two bodies considered here, namely the Earth and the moon. The main forces (omitted in figure 10 for the sake of clarity) are naturally the forces of gravitational mutual attraction, equal and opposite, applied to the centers of the Earth and the moon. The remaining two forces, labeled F_1 and F_2 in the figure, are those exerted by the moon on the bulges D_1 and D_2 of the oceans' surfaces. These forces are projected on the one hand, onto the Oy axis so as to reinforce the force of attraction on the Earth and, on the other hand, onto the Oz axis so as to provide a noncentral component F_{me} whose point of application is the center O and whose direction is that of the negative z's (inasmuch as $F_1 > F_2$). According to the generalized law of equal action and reaction, to this force F_{me} exerted by the moon on the Earth there must be an equal and opposite (noncentral) force F_{em} exerted by the Earth on the moon (see figure 10).

Increase in the Distance from the Earth to the Moon

This noncentral force F_{em} acts in the direction of the moon's motion in its circular orbit around the Earth. Paradoxically, this same force that "accelerates" the moon also results in a decrease in its velocity v_m and an increase in the radius r_m of its circular trajectory. Exactly the opposite occurs when artificial satellites orbiting the Earth enter the upper layers of the atmosphere: collisions between the satellite and the atoms or molecules "slow down" the space vehicle as its speed increases and its distance from the center of the Earth diminishes. The moon's movement away from the earth can thus be explained by the presence of this force F_{em}. The increase in the distance from the Earth to the moon per unit of time as determined by measurements and observations is equal to:

$$dr_m/dt = 3 \text{ cm/year} = 3 \text{ m/century}$$

(*Encyclopaedia Britannica*, 15th ed., s.v., "Tidal Friction").

In 1969 the Americans set up special mirrors on the surface of the moon. These mirrors are capable of reflecting light rays received in the incident direction. Since this date, a method of evaluating r_m has consisted in transmitting laser pulses to these mirrors from the Earth, then measuring the interval of time it takes for the transmitted pulse to be reflected back to Earth. Using the equation $\Delta t = 2\, r_m/c$, it has been possible to determine the Earth–moon distance.

The Slowing Down of the Earth's Rotation

When noncentral forces of interaction exist, the generalized law of equal action and reaction necessarily implies that couples are involved. In the Earth–moon example being considered here, the moment of the Earth–moon forces of interaction evaluated at 0 must be equal to and opposite the moment of the moon–Earth forces of interactions evaluated at the same point. However, figure 10 allows us to see that the moment of the noncentral force F_{em} evaluated at 0 is not null and is expressed: $M_{o\ (earth-moon)} = r_m\, F_{em}\, e_x$. A couple C_{me} must therefore be acting on the Earth, and its axis must be the Ox axis:

$$M_{o\ (moon-earth)} = C_{me} = -\,r_m\, F_{em}\, e_x,\ \text{in such a way that}$$
$$M_{o\ (earth-moon)} = -\,M_{o\ (moon-earth)}$$

In figure 10, we have represented this couple with axis Ox by the two equivalent forces $+\,f_c$ and $-\,f_c$ applied at two points diametrically opposite each other on the Earth's equator. This couple slows down the Earth's daily rotational motion. The phenomenon of the Earth's gradual slowing down has been studied in depth in numerous works, one of which is "Long-term Development in the Speed of the Earth's Rotation Obtained from the Orbital Motion of the Moon and the Planets" (in *Encyclopédie Scientifique de l'Univers, La Terre* [Paris: Bureau des Longitudes, Bordas, 1984]: 50). This study revealed that the average gradual lengthening of the day is:

$$dT/dt = 1.6.10^{-5}\ \text{s./year} = 1.6\ \text{ms/century}$$

(T = 24 hours = 86,400 seconds = the average length of one day.)

The Law of Conservation of Angular Momentum

The total angular momentum of the Earth–moon system as evaluated at point G, the center of mass of the same system, is constant in modulus and direction. This property is a direct consequence of the application of Newton's laws. As the angular momentum L_O of the same system, calculated at point O—the center of the Earth—differs from that measured at G only by a constant, we conclude that L_O must be constant in modulus and direction. Now the two main components of L_O are (1) the angular momentum $I_e \, \Omega$ of the Earth in its rotation on its polar axis and (2) the angular momentum $M_m r_m v_m$ of the moon in its circular orbit around the Earth. If the sum of these two components is constant, a decrease (slowing down) in the first results in an increase in the distance from the Earth to the moon. Mathematically, we have:

$$I_e \, \Omega \, \frac{dT}{T} = \tfrac{1}{2} \, m_m \, v_m \, r_m \, \frac{dr_m}{r_m}$$

In light of the numerical values of both constants and experimental values relative to $dT = 1.6$ ms and $dr_m = 3$m per century, calculations yield:

$$I_e \, \Omega \, \frac{dT}{T} = 1.32 \cdot 10^{26} \text{ and:}$$

$$\tfrac{1}{2} \, m_m \, v_m \, r_m \, \frac{dr_m}{r_m} = 1.13 \cdot 10^{26} \text{ kg. m}^2 \text{ s}^{-1}$$

The two numerical results thus obtained are sufficiently close to convince us that the two phenomena—a slowing down of the Earth's rotation and an increase in the distance from the Earth to the moon—are connected.

To conclude, this section has shown that Newton's generalized laws are closely linked to the laws of conservation of angular momentum and energy in an isolated system. We have performed the calculations concerning angular momenta. In assessing energy totals we find:

- a decrease in the kinetic energy of the Earth's rotation;
- a decrease in the kinetic energy of the moon's rotation;
- an increase in the Earth–moon gravitational interaction energy;
- finally, the dissipation in the form of heat due to friction between the Earth's crust and the oceans.

Conclusion

Relativity and Universality

For nearly two centuries the universality of Newton's laws has gone almost unchallenged. With the arrival of Albert Einstein's theories of special and general relativity the scientific world was shaken by doubt and confusion. The first criticisms and attacks seemed justified. Didn't classical mechanics state that the propagation speed of certain interactions had to be infinite? There followed attacks that often appeared less valid: the law of equal action and reaction could be violated! These criticisms need no longer exist if Newton's laws are interpreted correctly. In this book we have shown that when these laws are generalized they remain universal. In other words, Einstein may be right without Newton's being wrong.

This generalization, which is aimed at preserving the spirit of Newton's statements, is based on certain basic postulates of modern physics. In fact, the generalized statements we have offered are in perfect agreement with the laws of conservation of momentum, angular momentum, and energy in an isolated system. These very same (generalized) statements of Newton's laws take into account the fact that all interactions occurring between two bodies (on the atomic or astronomical scale) take place due to exchanges of momentum and/or angular momentum.

Newton's laws are universal: The law of gravitational attraction is obviously unchallenged. As for the three laws of motion, they are also universal, since they apply not only to particles, but also to solid *bodies* (whether liquid or gas), and because they are valid in any reference system whatsoever (Galilean or non-Galilean). A simultaneous application of Newton's three laws makes it possible to investigate the properties of the driving inertial forces and Coriolis forces and to state that these properties are similar to those of other "classical" forces (gravitational, electromagnetic, etc.). This is in perfect agreement with Einstein's equivalence principle, which states that gravitational and inertial masses are equal.

And finally, Newton's laws are universal to the extent that their applicability may be extended to relativistic speeds.

References

1. Whipple, F. L. 1980, *Scientific American* 242, no 3: 88.
2. Gié, H. 1965, *Dynamique*. Paris: J. B. Baillière et fils, p. 66.
3. Einstein, A., and L. Infeld. 1981, *L'évolution des idées en physique*. Paris: Payot.
4. Symon, K. R. 1965, *Mechanics*, 2d ed. Reading, Mass.: Addison-Wesley, p. 40.
5. Einstein, A. *La théorie de la relativité restreinte et gœnérale*. Paris: Gauthier-Villars.
6. Couderc, P. 1981, *La relativité*. Paris: P.U.F., p. 66.
7. Chazy, J. 1928, *La théorie de la relativité et la mécanique céleste*. Paris: Gauthier-Villars.
8. Purcell, E. M. 1965, *Electricity and magnetism*. New York: McGraw-Hill, pp. 173–87.